Synthetic Organic Chemistry and the Nobel Prize, Volume 2

The Nobel Prize is the highest award in science, as is the case with non-science fields too, and it is, therefore, arguably the most internationally recognized award in the world. This unique set of volumes focuses on summarizing the Nobel Prize within organic chemistry, as well as the specializations within this specialty. Any reader researching the history of the field of organic chemistry will be interested in this work. Furthermore, it serves as an outstanding resource for providing a better understanding of the circumstances that led to these amazing discoveries and what has happened as a result, in the years since.

T0353566

Synthetic Organic Chemistry and the Nobel Prize

Synthetic Organic Chemistry and the Nobel Prize Volume 1
John G. D'Angelo

Synthetic Organic Chemistry and the Nobel Prize Volume 2
John G. D'Angelo

Synthetic Organic Chemistry and the Nobel Prize, Volume 2

John G. D'Angelo

CRC Press
Taylor & Francis Group
Boca Raton London New York

CRC Press is an imprint of the
Taylor & Francis Group, an informa business

First edition published 2023
by CRC Press
6000 Broken Sound Parkway NW, Suite 300, Boca Raton, FL 33487-2742

and by CRC Press
4 Park Square, Milton Park, Abingdon, Oxon, OX14 4RN

Library of Congress Cataloging-in-Publication Data
Names: D'Angelo, John, author.
Title: Synthetic organic chemistry and the Nobel Prize / John G. D'Angelo.
Description: First edition. | Boca Raton : CRC Press, 2022. |
Series: Synthetic organic chemistry and the nobel prize | Includes bibliographical references and index.
Identifiers: LCCN 2022034953 (print) | LCCN 2022034954 (ebook) |
ISBN 9780367438975 (volume 1 ; hardback) | ISBN 9781032417202 (volume 1 ; paperback) | ISBN 9781003006831 (volume 1 ; ebook)
Subjects: LCSH: Organic compounds—Synthesis. | Nobel Prizes. |
Nobel Prize winners. | Chemistry, Organic—Awards.
Classification: LCC QD262 .D36 2022 (print) | LCC QD262 (ebook) |
DDC 547.0079—dc23/eng20221021
LC record available at https://lccn.loc.gov/2022034953
LC ebook record available at https://lccn.loc.gov/2022034954

ISBN: 978-0-367-43898-2 (hbk)
ISBN: 978-1-003-00684-8 (ebk)
ISBN: 978-1-032-51906-7 (pbk)

DOI: 10.1201/9781003006848

Typeset in Times
by codeMantra

Contents

vi Contents

About the author

Dr. John G. D'Angelo earned his BS in Chemistry from the State University of New York at Stony Brook in 2000. While at Stony Brook, he worked in Prof. Peter Tonge's lab on research toward elucidating the mechanism of action of FAS-II inhibitors for anti-mycobacterium tuberculosis drugs. He was also an active member of the chemistry club, serving as its treasurer for a year. After graduating, he worked as a summer research associate at Stony Brook in Prof. Nancy Goroff's lab, working toward the synthesis of molecular belts. He then earned his PhD from the University of Connecticut in 2005, working in the laboratories of Michael B. Smith. There, Dr. D'Angelo worked on the synthesis of 2-nucleobase, 5-hydroxymethyl lactams as putative anti-HIV agents while also investigating the usefulness of the conducting polymer poly-(3,4-ethylinedioxy thiophene) as a chemical reagent. He served as a teaching assistant during most of his five years at UCONN and was awarded the Outstanding TA award during one of these years.

After completing his PhD, he took a position as a postdoctoral research associate at The Johns Hopkins University in Prof. Gary H. Posner's lab. There, Dr. D'Angelo worked on the development of artemisinin derivatives as anti-malarial and anti-toxoplasma gondii derivatives. In 2007, Dr. D'Angelo accepted a position at Alfred University at the rank of Assistant Professor, and in 2013, he was awarded tenure and promotion to the rank of Associate Professor at Alfred and awarded promotion to Professor in July 2021. Dr. John G. D'Angelo's research continued for a while to focus on the chemical reactivity of conducting polymers and has been expanded to pedagogical research and scientific ethics; the latter two now the focus of his research efforts. He served as the local ACS section (Corning) chair in 2014 and 2021 and as the Faculty Senate president for two consecutive terms serving in this capacity from 2014 to 2018 and became Chair of the Chemistry Division at Alfred in 2021.

He is also the author of four books. One, on scientific misconduct, is in its second edition, and the second book on scientific misconduct is intended to be a workbook with hypothetical cases that students can work through. The third book, written with his PhD advisor, outlines a process for using the chemical search engine Reaxys to teach reactions, and the fourth book is a now discontinued organic chemistry textbook published through the web-based publisher Top Hat. He is also an author of 13 peer-reviewed publications (three in his independent career) and two patents. This four-volume series on organic chemistry and the Nobel Prize is his latest authoring endeavor.

Introduction

1

HISTORY

The Nobel Prize—a prize recognized at least in name—as one of, if not, THE premier reward for genius is arguably the most famous award in the world. It is unlikely that someone past a High School education has never heard of the Nobel Prize. Awarded (mostly) annually since 1901 in the subjects of Chemistry, Physics, Physiology or Medicine, Peace, and Literature and joined in 1968 by the Sveriges Riksbank Prize in Economic Sciences in the memory of Alfred Nobel, these prizes carry a medal, diploma, and cash prize for those chosen for this high honor in addition to the accompanying recognition. It is important to note that the Economic Sciences prize is formally speaking a not Nobel Prize, though they're awarded at the same time, and this prize is treated very much like a Nobel (Figure 1.1).

FIGURE 1.1 Artist rendition of Alfred Nobel (Image #1704216421).

DOI: 10.1201/9781003006848-1

The prize was created by Alfred Nobel in 1895 in his last will and testament, with the largest share of his considerable fortune allocated to the series of prizes. The "rules" set out in Nobel's will about the award are still at least mostly adhered to today, over a century later. The chief difference is that although Nobel stipulated that the award should be made for a scientific matter from the *preceding year*, it appears (to me anyway) that it has more recently become more of a lifetime achievement award of sorts, at least in the life and physical sciences awards (Chemistry, Physiology or Medicine, and Physics). Exactly when this started is difficult to pin down, but it is very clearly the current modus operandi.

The primary source of Nobel's wealth was being the inventor of dynamite, a stabilized form of the explosive nitroglycerine. In a very real way, the establishment of the prize was its own first controversy. Initially, the creation of the prize caused somewhat of a scandal, and it wasn't until several years (1901) after his death (1896) that his requests were finally fulfilled with the first series of awards. His own family opposed the prize; much of his considerable wealth had been bequeathed for its creation rather than to them so it is easy to understand their objection. In his will, he specifically called for the Swedish Academy of Sciences to award both the Physics and Chemistry awards; the Karolinska Institute in Stockholm to award the Physiology or Medicine award; the Peace prize to be awarded by a committee of five to be selected by the Norwegian Storting; and the literature award to be awarded by the Academy in Stockholm.

One may wonder why Alfred Nobel created the prize at all. Although there is no direct evidence to support the claims, legend has it that Alfred was horrified by an errant obituary—his own—mistakenly published upon the death of his *brother*. In it, Alfred was referred to as "the merchant of death," due to how his invention, dynamite, had been used. In addition to less violent uses (e.g. mining), dynamite and other modern derivatives/analogs are also used as a weapon and Nobel's time was no different. Being labeled a "merchant of death" is enough to rattle just about anyone's emotional cage. The timing (1895) of his will (his third one) seems to fit this legend since the version of his last will and testament establishing the award was signed seven years after the death of his brother, rather than before his brother's death while his wealth had come far earlier.

OTHER FACTS

The awards are traditionally announced over a one-week period in early October. The ceremony and lectures are held in Stockholm later in the year, usually in December. As of this writing, in mid-2022, since its inception the

Nobel Prize has been awarded to 947 individual Laureates and 28 organizations. Since great intellect and achievements often are not isolated accomplishments, four individuals have won more than one Nobel Prize and three organizations have done likewise. Furthermore, some Nobel greatness apparently "runs" in some families. Among these, the Curie family is by far the most successful though they are not alone in having more than one family member earn an award.

Although there are no rules limiting how many awards one recipient can win, no award can be given for more than two works (i.e., topics) in any given year and the award may only be shared, whether it is for one work or two, between three recipients among each of the awarded fields. In some cases, this sharing will happen by more than one topic earning the honor, other times, it will be due to more than one individual or organization contributing to the same work. In short, the following permutations are the only ones allowed.

- One topic, one person
- One topic two people
- One topic three people
- Topic A one person and Topic B one person
- Topic A one person and Topic B two people

Prior to 1974, posthumous awarding of the prize was permitted and happened twice. Dag Hammarskjöl (1961, Peace) and Erik Axel Karlfeldt (1931, Literature) were both posthumously awarded the Nobel Prize. Both died earlier in the year they were awarded their prize and likely were at the very least nominated prior to their deaths. Hammarskjöl, the second Secretary-General of the United Nations was given the award "for developing the UN into an effective and constructive international organization, capable of giving life to the principles and aims expressed in the UN Charter." Hammarskjöl died in a plane crash in September 1961. Karlfeldt, meanwhile, was given the award for "the poetry of Erik Axel Karlfeldt" and died in April 1931. One exception since 1974 has been made for extenuating circumstances. The committee did not know that the awardee, Ralph Steinman (2011, Physiology or Medicine) had passed away a mere three days before announcing the award. Steinman was given the award "for his discovery of the dendritic cell and its role in adaptive immunity."

The maximum number of recipients stipulated has become a modern source of controversy. Science, and in fact the world, is far more collaborative than it was in Nobel's time. Rarely—effectively never—does a modern researcher make a major discovery alone. Instead, most modern work involves literal scores of individuals playing an important even if small role in creating the final mosaic. Thus, restricting the Nobel Prize to such a small number guarantees that people contributing to the work are left out. It is even safe to say it leaves out *most* of the people contributing. That it is inevitably and arguably often the people doing the lion's share of actual work only magnifies the problem. This is a fair criticism of the prize. However, we must recognize that it is also fair to lay this claim against *any* award. Even if it is more noticeable because of the magnitude of *this* prize, all awards (including sports awards) that recognize an *individual* ignore the contributions of essential supporting role players. By no means am I trying to justify or defend this reality. I only wish to point out that all awards can be so criticized. This is covered in more detail, as are some potential resolutions, in the controversies section later in this introduction.

Nobel's will also stipulates what should happen in years where no award is given for a field. The will states, the prize money is to be reserved for the following year and if even then, no award can be made, the funds are added to the Foundation's restricted funds. This (no award for a field) has happened, between the awards, a total of 49 times, mostly during times of war. Technically, the statutes for skipping a year refer to it being possible to not make an award if "none of the works under consideration is found to be of the importance indicated in the first paragraph..." but that the award was consistently not awarded during times of global conflict is probably not a coincidence. It is unlikely that somehow, no important work was done during those years. Literature was skipped in 2018 amid controversy involving one of the committee members though in this case, it was awarded the next year, making delayed more accurate in this case than skipped. Peace is the one "skipped" the most at 19 times. The others do not even measure to half, in order: medicine—9; chemistry—8; literature—7; and physics—6. The award for economic sciences has not existed during any times of global conflict and has never been skipped.

(Perhaps) surprisingly, two Nobel Laureates declined the prize: Jean-Paul Sartre (1964, Literature) on the grounds that he consistently declined **all** official honors, and Le Duc Tho (1973, Peace) along with U.S. Secretary of State (who accepted the award), Henry Kissinger. Although they were jointly awarded the prize for their work on the Vietnam Peace Accord, Tho pointed to the ongoing situation in Vietnam as justification for declining. Four others were **forced** to decline the award. Three of the four were Germans—Richard Kuhn (1938, Chemistry) "for his work on carotenoids and vitamins," Adolf Butenandt (1939, Chemistry) "for his work on sex hormones," and Gerhard Domagk (1939, Physiology or Medicine) "for the discovery of the antibacterial effects of prontosil"—forbidden from accepting the award by Hitler; all four later were able to receive the diploma and medal but not the prize. The fourth, Boris Pasternak, a Russian, (1958, Literature) "for his important achievement both in contemporary lyrical poetry and the field of the great

Russian epic tradition," initially accepted his award but was later coerced by authorities to decline it.

There are also no restrictions regarding the awardee being a free person; three laureates were imprisoned at the time of the award. Carl von Ossietzky (1935, Peace), Aung San Suu Kyi (1991, Peace), and Liu Xiaobo (2010, Peace) were all awarded the Prize while incarcerated. von Ossietzky was given the award "for his burning love for freedom of thought and expression and his valuable contribution to the cause of peace" and was an anti-Nazi who revealed the rearmament efforts of Germany in violation of the Versailles Treaty that ended World War I. He was sent to a concentration camp when the Nazis seized power. Hitler's fury in response to von Ossietzky's award led him to prohibit all Germans from receiving the Nobel Prize. Kyi was awarded the prize "for her non-violent struggle for democracy and human rights." She opposed the military junta that ruled Burma, efforts that landed her under house arrest for nearly 15 years. After being released, she resumed her political career only to be arrested again after a military coup and later sentenced to a total of 8 years. Finally, Liu Xiaobo, given the award "for his long and non-violent struggle for fundamental human rights in China," received his sentence for the crime of speaking. His first stint in prison was due to his part in the student protests on Tiananmen Square in 1989 and a second (this time in a labor camp) for his criticism of China's one-party system. Most recently, in 2008, Liu co-authored Charta 08, which advocates for China's shift in the direction of democracy. His charge was undermining the state authorities, and this earned him an 11-year sentence.

Some Nobel Laureates were downright deplorable. Take for example Dr. D. Carleton Gajdusek (1976, Physiology or Medicine), a pediatrician who discovered the role of prions in a disease known as Kuru, which is related to mad cow disease. He was also a self-admitted, and I dare say unapologetic pedophile. Many of his victims were also his research patients. Another, Fritz Haber (1918, Chemistry) was potentially a war criminal. Both are covered in more detail in the controversies section of this introduction.

Sometimes, Nobel greatness runs in the family. The Curie Family is the most prolific of the "Nobel Families." Pierre Curie won the prize (1903, Physics), sharing it with his wife, Marie (a.k.a. Madame) Curie, who went on to win one of her own (1911, Chemistry) several years later. One of their daughters, Irène Joliot-Curie and her husband Frédèric Joliot also went on to share a Nobel Prize (1935, Chemistry). This brings their family total to five shared or individual Nobel Prizes. As if this were not enough, although not an actual awardee, Henry R. Labouisse, husband of another of Marie and Pierre's daughters Ève, accepted the prize on behalf of UNICEF (1965,

Peace). All told, this one family had a hand in no less than **six** Nobel Prizes. It is extremely unlikely that something like this will *ever* be matched. It appears that the Curies are the New York Yankees of the Nobel Prize. The Curies, though the most prolific, do not hold an exclusive claim to multiple family members earning a Nobel Prize. Other spousal pairs to share a prize are: Carl and Gerty Cori (1947, Physiology or Medicine); Esther Duflo and Abhijit Banerjee (2019, Sveriges Riksbank Prize in Economic Sciences in Memory of Alfred Nobel); and May-Britt and Edvard I. Moser (2014, Physiology or Medicine) all shared a Nobel Prize while Gunnar Mydral (1974, Sveriges Riksbank Prize in Economic Sciences in Memory of Alfred Nobel) and his spouse Alva Mydral (1982, Peace) brought two separate Nobels to their house. Father and son pairs have also brought home Nobel Prizes with one pair: Sir William Henry and son William Lawrence Bragg (1915, Physics) sharing a prize. Other father-son pairs include Niels (1922, Physics) and Agae Bohr (1975, Physics); Hons von Euler-Chelpin (1929, Chemistry) and Ulf von Euler (1970, Physiology or Medicine); Arthur (1959, Physiology or Medicine) and Roger Kornberg (2006, Chemistry); Manne (1924, Physics) and Kai Siegbahn (1981, Physics); and J. J. (1906, Physics) and George Thomson (1937, Physics). Rounding out the keeping it in the family trend are Jan (1969 Sveriges Riksbank Prize in Economic Sciences in Memory of Alfred Nobel), and younger brother Nikolaas Tinbergen (1973, Physiology or Medicine).

How is the Nobel Prize selected?

Each prize has a selection committee that, around September of the preceding year, sends confidential forms to individuals considered qualified and competent to nominate. Committee members are all members of the academy and serve for a period of three years. Not just anyone can serve as an expert advisor, as well; only those specially appointed.

Across the awards, nominations are not allowed to be revealed until 50 years after the prize has been awarded. However, nominators are under no obligation to keep their nominations confidential. The qualified nominators and timeline (which does have some slight variations) for each award are summarized below.

The last point across all the awards effectively serves as an "out" of sorts to allow for solicitation from anyone in the field. Importantly, nobody can self-nominate. One would imagine (or hope) such hubris is unlikely to be received well by the committee, anyway. Likewise, at least for the chemistry prize, it is not possible for *just anyone* to submit a nomination; for example, you cannot nominate one of your pals.

FAMILY	NAME(S)	YEAR	FIELD	TOPIC
	Pierre and Marie Curie	1903	Physics	"In recognition of the extraordinary services they have rendered by their joint researches on the radiation phenomena discovered by Professor Henri Becquerel"
	Marie Curie	1911	Chemistry	"In recognition of her services to the advancement of chemistry by the discovery of the elements radium and polonium, by the isolation of radium and the study of the nature and compounds of this remarkable element"
	Irène Joliot-Curie and Frédéric Joliot	1935	Chemistry	"In recognition of their synthesis of new radioactive elements"
Curie	Henry R. Labouisse (on behalf of UNICEF)	1965	Peace	"For its effort to enhance solidarity between nations and reduce the difference between rich and poor states"
Cori	Carl and Gerty Cori	1947	Physiology or Medicine	"For their discovery of the course of the catalytic conversion of glycogen"
Duflo and Banerjee	Esther Duflo and Abhijit Banerjee	2019	Economic Sciences	"For their experimental approach to alleviating global poverty"
Moser	May-Britt and Edvard I. Moser	2014	Physiology or Medicine	"For their discoveries of cells that constitute a positioning system in the brain"
Mydral	Gunner Mydral	1974	Economic Sciences	"For their pioneering work in the theory of money and economic fluctuations and for their penetrating analysis of the interdependence of economic, social and institutional phenomena"
	Alva Mydral	1982	Peace	"For their work for disarmament and nuclear and weapon-free zones"
Bragg	Sir William and William Lawrence Bragg	1915	Physics	"For their services in the analysis of crystal structure by means of X-rays"

(Continued)

(Continued)

FAMILY	NAME(S)	YEAR	FIELD	TOPIC
Bohr	Niels	1922	Physics	"For his services in the investigation of the structure of atoms and of the radiation emanating from them"
	Agae	1975	Physics	"For the discovery of the connection between collective motion and particle motion in atomic nuclei and the development of the theory of the structure of the atomic nucleus based on this connection"
Euler-Chelpin	Hons von Euler-Chelpin	1929	Chemistry	"For their investigations on the fermentation of sugar and fermentative enzymes"
	Ulf von Euler	1970	Physiology or Medicine	"For their discoveries concerning the humoral transmitters in the nerve terminals and the mechanism for their storage, release and inactivation"
Kornberg	Arthur Kornberg	1959	Physiology or Medicine	"For their discovery of the mechanisms in the biological synthesis of ribonucleic acid and deoxyribonucleic acid"
	Roger Kornberg	2006	Chemistry	"For his studies of the molecular basis of eukaryotic transcription"
Siegbahn	Manne Siegbahn	1924	Physics	"For his discoveries and research in the field of X-ray spectroscopy"
	Kai Siegbahn	1981	Physics	"For his contribution to the development of high-resolution electron spectroscopy"
Thomson	J. J. Thomson	1906	Physics	"In recognition of the great merits of his theoretical and experimental investigations on the conduction of electricity by gases"
	George Thomson	1937	Physics	"For their experimental discovery of the diffraction of electrons by crystals"
Tinbergen	Jan Tinbergen	1969	Economic Sciences	"For having developed and applied dynamic models for the analysis of economic processes"
	Nikolaas Tinbergen	1973	Physiology or Medicine	"For their discoveries concerning organization and elicitation of individual and social behavior patterns"

CHEMISTRY[1]

QUALIFIED NOMINATORS	TIMELINE
Member of Royal Swedish Academy of Sciences Member of the Nobel Committee for chemistry or physics Nobel Laureate in chemistry or physics Permanent professor in the sciences of chemistry at the universities and institutes of technology of Sweden, Denmark, Finland, Iceland, and Norway and Karolinska Instituet, Stockholm Holders of corresponding Chairs in at least six universities or university colleges selected by the Academy of Sciences with a view to ensuring appropriate distribution of the different countries and their centers of learning Other scientists from whom the academy may see fit to invite proposals.	~September previous year: nomination invitations sent out. January 31: deadline for nominations March–June: consultation with experts June–August: report writing with recommendations September: academy receives the final report October: majority vote and announcement December: ceremony

PHYSICS[2]

QUALIFIED NOMINATORS	TIMELINE
Swedish and foreign members of the Royal Swedish Academy of Sciences; Members of the Nobel Committee for Physics; Nobel Prize laureates in physics; Tenured professors in the Physical sciences at the universities and institutes of technology of Sweden, Denmark, Finland, Iceland and Norway, and Karolinska Instituet, Stockholm; Holders of corresponding chairs in at least six universities or university colleges (normally, hundreds of universities) selected by the Academy of Sciences with a view to ensuring the appropriate distribution over the different countries and their seats of learning; and Other scientists from whom the Academy may see fit to invite proposals.	~September previous year: nomination invitations sent out. January 31: deadline for nominations March–June: consultation with experts June–August: report writing with recommendations September: Academy receives the final report October: majority vote and announcement December: ceremony

Why Laureates?

The reference to Nobel Prize winners as laureates has its roots in ancient Greece. A laurel wreath, a circular crown made of branches and leaves of the bay laurel, was awarded to victors of athletic competitions and poetic meets

PHYSIOLOGY OR MEDICINE[3]

QUALIFIED NOMINATORS	TIMELINE
Members of the Nobel Assembly at Karolinska Instituet, Stockholm; Swedish and foreign members of the Medicine and Biology classes of the Royal Swedish Academy of Sciences; Nobel Prize laureates in physiology or medicine and chemistry; Members of the Nobel Committee not qualified under paragraph 1 above; Holders of established posts as full professors at the faculties of medicine in Sweden and holders of similar posts at the faculties of medicine or similar institutions in Denmark, Finland, Iceland, and Norway; Holders of similar posts at no fewer than six other faculties of medicine at universities around the world, selected by the Nobel Assembly, with a view to ensuring the appropriate distribution of the task among various countries. Scientists whom the Nobel Assembly may otherwise see fit to approach. No self-nominations are considered.	~September previous year: nomination invitations sent out. January 31: deadline for nominations March–June: consultation with experts June–August: report writing with recommendations September: academy receives the final report October: majority vote and announcement December: ceremony

LITERATURE[4]

QUALIFIED NOMINATORS	TIMELINE
Members of the Swedish Academy and of other academies, institutions, and societies which are similar to it in construction and purpose; Professors of literature and linguistics at universities and university colleges; Previous Nobel Prize laureates in literature; Presidents of those societies of authors who are representative of the literary production in their respective countries.	~September previous year: nominations sent out January 31: deadline for nominations April: 15–20 preliminary candidates May: five final candidates June–August: reading of productions September: academy members confer October: award announced December: ceremony

PEACE[5]

QUALIFIED NOMINATORS	TIMELINE
Members of national assemblies and national governments (cabinet members/ministers) of sovereign states as well as current heads of states Members of The International Court of Justice in The Hague and The Permanent Court of Arbitration in The Hague Members of l'Institut de Droit International Members of the international board of the Women's International League for Peace and Freedom University professors, professors emeriti and associate professors of history, social sciences, law, philosophy, theology, and religion; university rectors and university directors (or their equivalents); directors of peace research institutes and foreign policy institutes Persons who have been awarded the Nobel Peace Prize Members of the main board of directors or its equivalent of organizations that have been awarded the Nobel Peace Prize Current and former members of the Norwegian Nobel Committee (proposals by current members of the Committee to be submitted no later than at the first meeting of the Committee after 1 February) Former advisers to the Norwegian Nobel Committee	~September previous year: nomination invitations sent out. January 31: deadline for nominations February–April: preparation of short list April–August: adviser review October: majority vote and announcement December: ceremony

ECONOMIC SCIENCES[6]

QUALIFIED NOMINATORS	TIMELINE
Swedish and foreign members of the Royal Swedish Academy of Sciences; Members of the Prize Committee for the Sveriges Riksbank Prize in Economic Sciences in Memory of Alfred Nobel; Persons who have been awarded the Sveriges Riksbank Prize in Economic Sciences in Memory of Alfred Nobel; Permanent professors in relevant subjects at the universities and colleges in Sweden, Denmark, Finland, Iceland, and Norway; Holders of corresponding chairs in at least six universities or colleges, selected for the relevant year by the Academy of Sciences with a view to ensuring the appropriate distribution between different countries and their seats of learning; and Other scientists from whom the Academy may see fit to invite proposals.	~September previous year: nomination invitations sent out. January 31: deadline for nominations March–June: consultation with experts June–August: report and recommendation writing September: Academy receives report on finalists October: majority vote and announcement December: ceremony

FIGURE 1.2   Laurel Wreath (IMG # 448716724).

in ancient Greece as a sign of honor. The term laureate to describe awardees of the Nobel Prize is used to harken back to this honor (Figure 1.2).

The medal, diploma, and cash prize

Created by Swedish and Norwegian artists and calligraphers each diploma is quite literally a work of art. The medals all have Nobel's image on them with his birth and death years, though the economic sciences prize is slightly different in its design. Recall that this prize is in Nobel's memory and was not one of the original subjects endowed by Nobel. Each medal is hand-made out of 18-carat recycled gold. The cash award is currently set at nine million Swedish Krona. As of this writing, this is equal to ~851,000 euros and ~965,000 USD (Figure 1.3).

Nobels proven wrong?

It is important to remember that science very routinely corrects itself. The mistakes that are corrected sometimes are because of misconduct, but other times, it is because technologies or other understandings improve, and we learn that the initial conclusions were wrong. It is very difficult to

FIGURE 1.3 Nobel Medal (IMG #1652139046).

impossible to say which *really* happens more often, but I sense misconduct is less common. It should be no surprise that work that earned someone a Nobel Prize is not immune to this fate. To date, no Nobel Prizes have been awarded for work later found to be fraudulent. Two examples of *disproven* works are Johannes Fibiger (1926, Physiology or Medicine) and Enrico Fermi (1938, Physics). Fibiger was awarded the Nobel for the discovery of a parasitic cancer-causing worm. Although the worm without question is real, subsequent research proved there were no cancer-causing properties of this parasitic infection. Fibiger died (1928) before his work could be proven wrong.

Fermi, on the other hand, lived long enough to see his Nobel-winning work disproven and even agreed with the new (and correct) conclusions. Fermi had incorrectly concluded he generated new chemical elements during some of his nuclear chemistry work. He even went so far as to bestow names upon them. What Fermi had actually done, and Otto Hahn eventually figured out, was cause nuclear fission to occur. That this phenomenon was never documented before without question led to Fermi's incorrect conclusion. It simply could not—at the time—have been the predicted outcome. Only later did additional work shed light on what was really happening.

Other award-winning works seem dubious, at best, with modern hindsight. Paul Hermann Müller (1948, Physiology or Medicine) was awarded the Nobel Prize for his discovery of the anti-insect properties of DDT, a substance that however effective at controlling mosquitos—and as a result malaria suppression—is now banned globally except in very exceptional circumstances. Its environmental impacts were later found to be severe enough that DDT is one of the central topics in Rachel Carson's "Silent Spring," a work credited with igniting the environmental movement. Carlson died relatively young at 57 (in 1964). Perhaps, she would have been a Nobel Laureate (for Peace) for her conservation work had she not. Finally, Antonio Egas Monic (1949, Physiology or Medicine) earned his award for developing the medical procedure lobotomy, a procedure rarely used today.

Nazis and the disappearing Nobel Prizes

Something you are unlikely to find in your standard world history text is that two Nobel Prizes were dissolved to keep them from being captured by the Nazis as they over-ran Europe. German scientists, Max von Laue (1914, Physics, for his discovery of diffraction of X-rays by crystals) and James Franck (1925, Physics, with Hertz, for his discovery of the laws governing the impact of an electron on an atom), sent their Nobel Medals out of Germany to fellow physicist Niels Bohr in Copenhagen for safekeeping. If caught doing this, von Laue and Franck likely would have been executed. Unfortunately, before long Copenhagen was likewise no longer safe; the Nazis were high stepping through the streets. An associate of Bohr's, George de Hevesy, made the bold decision to hide the medals by *dissolving* them. That is right, he dissolved the gold medals much like any of us dissolve sugar in a cup of coffee or a spot of tea. Gold is rather inert though, so dissolving it is difficult, requiring a solution that sounds like a bad idea to EVER make to the casual reader: a 3:1 mixture of hydrochloric acid:nitric acid. The result of this bold maneuver was two beakers of an orange solution. These beakers were set on a high shelf, and de Hevesy eventually left the beakers behind as he fled for Sweden; as a Jewish scientist, he was not what you would call safe in Nazi-controlled land. After the defeat of the Nazis, De Hevesy retuned to the lab. Miraculously (or perhaps it was the ghost of Alfred Nobel or maybe just luck) the beakers remained untouched. A little bit of chemistry later, the gold precipitated out and was sent to Stockholm. The Swedish Academy had the medals recast, *from their gold*, and the medals were (re-)presented to their rightful owners in a 1952 ceremony. Personally, I find this story to be a wonderful thumb to the eye of arguably the most terrible scourge in human history poked by science.

Controversies and snubs

Controversies

With honors such as the Nobel Prize, it should not be a surprise that there are occasional controversies, perceived snubs, or other general complaints and displeasure taken with both who is awarded the prize and maybe even more so who is not. There has even been some wondering whether or not the Nobel Prize is good for science at all.[7] Although a complete treatment of these is not appropriate in this book series—it could likely be a book on its own—the interested reader is encouraged to perform a simple internet search for "Nobel Prize controversies." While the issue is far more commonly centered around the work or one of the awardees, entirely unrelated to the **work** of a Nobel Prize, in 2018, the Nobel Prize for Literature was delayed until 2019 because of sexual harassment claims made against one of the board members. A brief discussion of some of the biggest work-based controversies is appropriate here.

There have been several Nobel Prizes that have been criticized in the sense that many believe the awardee should not have received the award. The Peace Prize seems particularly vulnerable to this. Former U.S. President Barak Obama (2009, Peace) is one of the most controversial. This is primarily because he was given the award a mere 9 months into his presidency. Nominations are collected *months* ahead of the actual award. This means that (then) President Obama, given the award "for his extraordinary efforts to strengthen international diplomacy and cooperation between peoples" was nominated incredibly early in his presidency; almost upon becoming the President. The notion that he, or frankly anyone, could have done enough to be worthy of this award is almost laughable.

Another controversial Peace Prize was given to Yasser Arafat, who shared the award with then-Israeli Prime Minister Yitzhak Rabin and Israeli Foreign Minister Shimon Peres (1994, Peace) because of their collective work on the Oslo Peace Accords. Not only were these Peace Accords unsuccessful at fostering peace—calling into question all three of the awards—Arafat was a key figure in many armed attacks in Israel; an awkward, at best, corollary.

Former U.S. Secretary of State Henry Kissinger's award (1973, Peace) was also very controversial. Recall his Vietnamese counterpart Le Duc Tho refused to accept his share of the prize, citing that there was still conflict. There was also the bombing of Hanoi, ordered by Kissinger, during cease-fire negotiations but before the award was given.

Perhaps, the most scandalous of all is Fritz Haber (1918, Chemistry). Fritz Haber is the Haber of the Haber–Bosch process, a process still used largely unchanged today to make ammonia, essential to the commercial production of fertilizer; this is why he won the Nobel. Because of this process, hundreds of millions, if not billions of people are fed. **This is not hyperbole**. Modern food production is possible because of his work. Whatever environmentalists want to (perhaps rightly) say about how sustainable this is and what the ceiling is, it is currently true. Just a few years *before* winning the prize, however, he orchestrated the first massive and deliberate chemical weapons attack during war, using chlorine gas against the allied forces in France during World War I.

D. Carleton Gajdusek (1976 Physiology or Medicine) is radically different from Fritz Haber. While Haber's crimes were done and well known *before* winning the prize, Gajdusek was not found to be a pedophile until years after his award. This means that Gajdusek's award was given ignorant of his vile actions while Haber was awarded his Nobel with full knowledge of the atrocities he committed. By no means am I comparing the crimes and ranking their "terribleness." There is, in my opinion, a stark difference between "holding your nose and making a decision" (Haber) and "later wincing at your decision" (Gajdusek) as disgusting information comes to light.

Nobel snubs

An extension of controversies is snubbing deserving winners. Before that, however, there are a number of giants of chemistry and science that could not have possibly been considered for the Nobel Prize, despite making contributions virtually unmatched in their importance. These pillars of chemistry, like those who solved the gas laws (Boyle, Charles, Gay-Lussac), Avogadro, and Lavoisier all predated Alfred Nobel's founding of the Prize. In fact, some predated Nobel himself. Lavoisier is arguably the most impactful among them. In addition to demonstrating the importance of oxygen to combustion—helping to finally vanquish phlogiston—Lavoisier's law of conservation of mass is perhaps the single most important contribution to chemistry *ever*. Thereafter, chemistry was a quantitative, rather than qualitative endeavor. It became a science according to more modern interpretations of science. Unfortunately for chemistry and especially Lavoisier, he was executed by guillotine at 50 during the French Revolution. In the broader science sense, Charles Darwin died (much less violently than Lavoisier) in 1882, well before the creation of the Nobel Prize, rendering him ineligible. It should also be further noted that Darwin's work does not exactly fit into any of the areas recognized by the Nobel Prize. Physiology or Medicine is the closest match and would be quite the stretch.

Such early scientists notwithstanding, it should come as no surprise—go ahead and be disappointed though—that the selection of the winner of the Nobel Prize is subject to the flaws of human behavior. By this, I mean that inevitably, biases, discrimination, and personal conflict have at least appeared to play a role in denying some otherwise worthy work being overlooked.

Even divorced from various biases, every year, it can be argued that "someone else" should have won the Nobel Prize, and this is probably true in every field, not just chemistry. In fact, this is the case for just about every award given across a very wide spectrum of recognition. Although sometimes the overlooked scientist eventually wins, the history of the Nobel Prize is littered with persons that it can be fairly asked "How did *they* never win?" Chemistry is not without such controversies. Also, multiple women have been snubbed a Nobel Prize in various fields. These snubs are covered in the appropriate section of this introduction. Some examples of snubs include the following:

Gilbert Lewis could have been awarded the Nobel Prize in Chemistry for more than one important discovery. Perhaps, his most noteworthy was the nature of a bond being the sharing of an electron pair. Lewis was nominated for the prize many times but was never awarded one. Reasons for him not winning one are difficult to prove, but there is some evidence that personal conflicts or less than fully informed opinions on the part of some evaluators played a significant role.

Dimitri Mendeleev, the father of the periodic table, was also never awarded a Nobel Prize in Chemistry. He allegedly came close, but the academy overseeing the award overruled the initial vote and changed the committee, subsequently holding a new vote that Mendeleev lost. Allegedly, this may have been at the behest of a rival (who at the time was a member of the Swedish Academy) of Mendeleev.

Henry Louis Le Châtelier was nominated for the prize but never awarded one despite his tremendously impactful contributions to chemical equilibria. It is possible that Le Châtelier's snub is a product of the historical fact that his most important work on equilibria was done too early (before the Nobel Prize was created), causing him to be overlooked because of the clause of "the previous year."

Chemistry and the sciences are not alone. Mahatma Gandhi, nominated five times for the Nobel Peace Prize, never won. Stephen Hawking, one of the most brilliant minds of our age was also never awarded a Nobel Prize for his work nor was Carl Sagan. This may have some root in that Hawking's work could not be tested with contemporary tools, but maybe not.

WOMEN AND THE NOBEL

Over the years, across all the disciplines, the prize has been awarded to a woman 58 times, with one, Marie Curie, winning a total of two, in different fields (Physics with her husband Pierre in 1903 and Chemistry, alone, in

1911). The physics prize has been awarded to women four times (2020, 2018, 1963, and 1903), chemistry seven times (2020 [two winners] 2018, 2009, 1964, 1935, and 1911), physiology or medicine 12 times (2015, 2014, 2009 [two winners], 2008, 2004, 1995, 1998, 1986, 1983, 1977, and 1947), literature 16 times (2020, 2018, 2015, 2013, 2009, 2007, 2004, 1996, 1993, 1991, 1966, 1945, 1938, 1928, 1926, and 1906), peace, which leads the way with 18 times (2021, 2018, 2014, 2011 [three winners], 2004, 2003, 1997, 1992, 1991, 1982, 1979, 1976, 1946, 1931, and 1905) and the economic sciences prize has been awarded to a woman twice (2019 and 2009).

AWARD	CHEMISTRY	PHYSIOLOGY AND MEDICINE	PHYSICS	PEACE	LITERATURE	ECONOMIC SCIENCES
Total	188	224	219	107	118	89
Women	7	12	4	18	16	2
% Women	4	5	2	17	14	2

To date, all Nobel Laureates have either outwardly presented as male or female. Even a casual perusal of the data should make clear that there is a wild imbalance with respect to the gender of the Nobel Prize winners. Currently, there has never been a winner who has announced being trans, and none of them transitioned later in their lives after being awarded a Nobel Prize.

Though earlier on, the dearth of women recipients in especially the science fields could be explained away by a smaller number of women in the fields, compared with men, modern women win rates are nowhere near the averages reported by employers and (anecdotally, at least) observed by anyone paying attention to the world. It is difficult to pin down a plausible, data or merit-based reason for this. Consequently, minds are left to wonder and inevitably wander to concluding discrimination and unfair judging. Although I want to say this is unjustified, I have no evidence to back up my stance. This is especially true for the awards with a more closed nomination process. Until and unless there is better accountability (perhaps through more transparent nominations) or systematic data-based rubrics that generate nominations or awards, this pall is likely to hang over the prizes.

Commentary on the dearth of women person of color laureates, especially in the sciences has been intensifying.[8] Although a shortage of nominees, at least for women has been cited,[9] concrete reasons for this lacking are far from agreed upon, a shortage of women in the fields is demonstrably false given that employment data indicate that the percentage of women working in the fields is higher than the percentage of women taking home Nobel Prizes. Even considering the unofficial transition to more of a lifetime achievement award, rather than the greatest achievement of the last year award does not hold water; women have been working in the sciences for decades. A list[10] has been compiled of deserving women chemists who arguably have been snubbed a Nobel Prize.

It would be quite easy to author an entire volume on a list of snubbed female Nobel-worthy scientists. The hardest part about such a list is determining when to stop. A short and nowhere near exhaustive list is:

Liese Meitner was one of the central contributors to nuclear fission, correcting Fermi's work along with Otto Hahn. Despite being a long-time collaborator of Hahn's, including on his award-winning work, Meitner did not share his award (1944, Chemistry). It is exceedingly difficult for anyone to justifiably rationalize this oversight even if some of her work was hampered as she fled for her life (Meitner was Jewish) to Sweden with the rise of the Nazis to power.

Rosalind Franklin (one of the most infamous snubs) could fairly consider to actually be not snubbed in the sense that the award was given for the work she contributed after her death. No sane and rational person can claim the committee waited for Franklin to die before recognizing the work. Nevertheless, it is impossible to ignore the fact that not one, not two, but three men were given the award related to her work. Part of what further enhances the ire of many is that particularly Watson and Crick collected little to no experimental results of their own regarding solving the structure of DNA, using especially Franklin's data on the way to do so. I for one believe—with admittedly no evidence—that she would have been awarded the prize had she not died. I see it happening in two different ways. First, Franklin could have been awarded the prize instead of Wilkins. This in my opinion is the most likely course as her data was viewed to be of higher quality than his. A second option would have been to make two different awards whereby Watson and Crick would be given one (likely Physiology or Medicine, though perhaps, Chemistry), and the other pair would be given the Physics or Chemistry prize.

Eleanor Roosevelt was an ardent advocate for Civil Rights and was never awarded the Peace Prize for her brave and noble work. Other Civil Rights activists (e.g., Rev. Dr. Martin Luther King, Jr.) were awarded the prize, so a claim that Civil Rights advocacy is insufficient to earn the award is weak, at best.

Joselyn Bell Burnell who performed the work that led to the discovery of pulsars sat by and watched while her colleagues won the Nobel Prize (1974, Physics). At the time, Burnell was a postdoctoral research student, further highlighting the disparity in credit between the Principal Investigators (PI) and those who work under their tutelage or mentoring.

Rachel Carlson's work is a cornerstone in the environmental movement. Had she not passed away in 1964, she likely would have been awarded a Novel Prize in Peace for her impact on the environmental movement. Instead, it wasn't until 43 years later that All Gore and the UN's Intergovernmental Panel on Climate Change were awarded the Nobel Prize for Peace for their work on increasing awareness regarding climate change.

Race and the Nobel Prize

The lack of racial diversity, particularly as it is measured by skin color, is as shocking as it is disappointing. Consider the table herein that summarizes the data.

PRIZE	CHEMISTRY	PHYSIOLOGY OR MEDICINE	PHYSICS	PEACE	LITERATURE	ECONOMIC SCIENCES
Total prizes	188	224	219	107	118	89
# Caucasian winners	173	214	195	80	102	86
% Caucasian	92	96	89	75	86	97
# Birth countries	38	38	35	47	49	19
# Affiliated countries	20	24	21	44	35	10

In most of the prizes, the numbers of non-Caucasian awardees are approximately the same level as women. Here, however, it may be easier to rationalize, but I want to fall well short of *justifying* it. This rationalization comes from the reality that at least in the cases of the research-based awards, the vast majority—in fact nearly all the winners—are affiliated with institutions within affluent countries that have sometimes invested billions of dollars in basic research annually. This higher level of funding inevitably leads to higher profile and higher quality—or at least higher sophistication—research. This is not to say that non-Caucasian researchers are, for certain, never overlooked. Once again, it appears inevitable that a black or brown person will be working at a high-profile and well-funded institution with vibrant research support. I even know more than a few such persons. As with women, there once again appears to be room to conclude only discrimination can produce this reality.

It is likely that an entire book can be written about the issue of general equality (be it gender, racial, or any other kind) in not just the Nobel Prizes but all manner of awards. One take that I heartily disagree with is the claim that it is a reflection of racism in the American education system.[11] Although the American education system is certainly rife with *inequality* that has many diverse roots, to claim that the American education system is at fault for a lack of diversity in a decidedly *international* award is absolutely off base. This may be part of why there is a dearth of racially diverse U.S.-based Nobel winners, it cannot possibly explain the issue on a global level. In fact, isolating the explanation to the education system in any country is potentially more counterproductive than it is helpful.

Outlined here is a breakdown of the country of origin and affiliation for all winners of each Nobel Prize category.

The chemistry award is spread over 18 individual affiliation countries and twice was awarded to someone who had a professional affiliation with an institution in two different countries. By more than a factor of two, the United States leads the way with a whopping 82 winners. The chemistry award by birth country is a bit more diverse, covering 38 different countries of origin for the awardee. Again, the United States is the leader by a very wide margin. The dominance of the United States in the Chemistry Prize (and other awards) begins shortly after World War II. Prior to that, western Europe was the leader. A similar observation can be made (at least with respect to the rise of the dominance by the United States) in other fields as well.

The Physiology or Medicine award has been awarded to individuals affiliated with 24 different countries, hailing from 38 different countries. Once again, the United States dominates the field. As was seen with the birth countries of some of the Chemistry Laureates, here we see countries that are no longer on the map. We also see how countries are referred to (e.g., Russian Empire vs. Russia and elsewhere USSR) changing. This is deliberately done to try to provide the best possible historical context of the number of awards. How countries were known during the award gives readers insight into the timeline and geopolitical climate of the respective eras.

The physics award has been awarded to individuals affiliated with 19 different countries with two awards being given to someone with multiple affiliations. As was the case in both the Chemistry and Physiology or Medicine awards, the number of birth countries ticks higher at 34. This is at least in part due to the ceasing to exist for some of these countries. Another part, however, is that people—including research scientists—relocate to other countries. In fact, after World War II, the United States "imported" (some may use the word poached) many topflight physicists from former enemy countries such as Germany.

Peace, the prize with the greatest diversity of all kinds, is one of the two award areas where the United States does not dominate. Part of why is that the leader is not any one individual but an organization. I have

CHEMISTRY

By affiliation

U.S.—82	Germany—33	U.K.—29	France—9	Switzerland—6
Japan—6	Sweden—5	Israel—4	Canada—3	Argentina—1
Austria—1	Belgium and U.S.—1	Czechoslovakia—1	Denmark—1	Finland—1
Italy—1	Netherlands—1	Norway—1	Switzerland and U.S.—1	
USSR—1				

By birth

U.S.—55	U.K.—25	Germany—24	France—11	Japan—7
Austria-Hungary—6	Prussia—5	Canada—4	Netherlands—4	Sweden—4
Austria—3	Russia—3	Russian Empire—3	Scotland—3	Switzerland—3
British Mandate of Palestine—2		Egypt—2	Hungary—2	Norway—2
Palestine—2	Australia—1	Austrian Empire—1	Bavaria—1	Belgium—1
China—1	Denmark—1	India—1	Italy—1	Korea—1
Lithuania—1	Mexico—1	New Zealand—1	Poland—1	Romania—1
South Africa—1	Taiwan—1	Turkey—1	West Germany—1	

PHYSIOLOGY OR MEDICINE

By affiliation

U.S.—115	U.K.—32	Germany—15	France—10	Switzerland—8
Sweden—7	Australia—4	Austria—4	Belgium—4	Denmark—4
Canada—3	Italy—3	Japan—3	Norway—2	Argentina—1
China—1	Dutch East Indies—1		Hungary—1	Netherlands—1
Portugal—1	Prussia—1	Russia—1	Spain—1	Tunisia—1

By birth

U.S.—80	U.K.—25	Germany—20	France—13	Australia—7
Sweden—6	Switzerland—6	Austria—5	Italy—5	Japan—5
Canada—4	Denmark—4	Austria-Hungary—3	Belgium—3	Netherlands—3
Russian Empire—3	Scotland—3	South Africa—3	Argentina—2	China—2
Norway—2	Poland—2	Spain—2	Austrian Empire – 1	Brazil—1
Hungary—1	Iceland—1	India—1	Lebanon—1	Luxembourg—1
Mecklenburg—1	New Zealand—1	Portugal—1	Prussia—1	Romania—1
Russia—1	Venezuela—1	Wurttemberg—1		

PHYSICS

Affiliation

U.S.—106	U.K.—27	Germany—20	France—13	Switzerland—9
Japan—7	Netherlands—7	USSR—7	Sweden—4	Canada—3
Denmark—3	Italy—3	Russia—2	Australia—1	Austria—1
Belgium—1	China—1	Germany and U.S.—1	India—1	Ireland—1
U.S. and Japan—1				

By birth

U.S.—71	Germany—24	U.K.—23	Japan—12	Netherlands—9
France—8	Canada—6	Italy—6	Russia—6	Switzerland—6
China—5	Prussia—4	Sweden—4	West Germany—4	India—3
USSR—3	Australia—2	Austria-Hungary—2	Austria—2	Denmark—2
Hungary—2	Poland—2	Russian Empire—2	Belgium—1	Czechoslovakia—1
French Algeria—1	Hesse-Kassel—1	Ireland—1	Luxembourg—1	Morocco—1
Norway—1	Russian Empire—1	Schleswig—1	Scotland—1	

PEACE

Affiliation/residence

Organization—27	U.S.—21	U.K.—12	France—9	Sweden—5
Germany—4	South Africa—4	Switzerland—4	Belgium—3	India—3
Ireland—3	Argentina—2	Austria—2	Canada—2	East Timor—2
Egypt—2	Israel—2	Liberia—2	Norway—2	USSR—2
Bangladesh—1	Burma—1	China—1	Colombia—1	Costa Rica—1
Democratic Republic of Congo—1		Denmark—1	Ethiopia—1	Finland—1
Ghana—1	Guatemala—1	Iran—1	Iraq—1	Italy—1
Japan—1	Kenya—1	Netherlands—1	Palestine—1	Philippines—1
Poland—1	Russia—1	South Korea—1	Vietnam—1	Yemen—1

Birth

U.S.—19	France—8	UK—8	Germany—6	Sweden—5
South Africa—4	Switzerland—4	Belgium—3	Egypt—3	Ireland—3
Argentina—2	East Timor—2	Liberia—2	Norway—2	Poland—2
Russian Empire—2	Scotland—2	USSR—2	Austria—1	Austrian Empire—1
British India—1	Burma—1	Canada—1	China—1	Colombia—1
Costa Rica—1	Denmark—1	Ethiopia—1	Ethiopian Congo—1	Finland—1
Gold Coast—1	Guatemala—1	India—1	Iran—1	Iraq—1
Japan—1	Kenya—1	Netherlands—1	Ottoman Empire—1	
Pakistan—1	Philippines—1	Romania—1	Russia—1	South Korea—1
Tibet—1	Vietnam—1	Yemen—1		

LITERATURE

By affiliation/residence

France—17	U.K.—13	U.S.—12	Switzerland—12	Germany—8
Sweden—8	Italy—6	Spain—5	Poland—4	Denmark—3
Ireland – 3	Norway – 3	USSR—3	Austria—2	Chile—2
Greece—2	Japan—2	Mexico—2	South Africa—2	Belgium—1
Canada—1	Czechoslovakia—1	Egypt—1	Finland—1	Guatemala—1
Hungary—1	India—1	Israel—1	Nigeria—1	Peru and Spain—1
Poland and U.S.—1	Portugal—1	St. Lucia—1	Turkey—1	Yugoslavia—1

By birth

France—11	U.S.—10	Germany—7	Sweden—7	Spain—6
U.K. – 6	Italy—5	Denmark—4	Ireland—4	Russia—4
Russian Empire—4	Japan—3	Poland—3	Austria—2	Austria-Hungary—2
Canada—2	Chile—2	China—2	India—2	Norway—2
South Africa—2	Belgium—1	Bosnia—1	Bulgaria—1	Crete—1
Colombia—1	Egypt—1	French Algeria—1	Guadeloupe Island—1	
Hungary—1	Iceland—1	Madagascar—1	Mexico—1	Nigeria—1
Ottoman Empire—1		Persia—1	Peru—1	Portugal—1
Prussia—1	Romania—1	Scheswig—1	St. Lucia—1	Switzerland—1
Tanzania—1	Trinidad and Tobago—1		Turkey—1	Tuscany—1
Ukraine—1	USSR – 1			

chosen to not try to parse out the country of origin of these organizations since it seems more important to me that the award went to an organization, rather than an individual compared to that organization's country of origin. As far as individuals go, however, the United States retains its spot at the top, even if by a slimmer than used to margin. The award has been given to individuals affiliated with or living in 43 different countries at the time of the award with a birth country number of 47. Of the Prizes considered so far—and literature will have this phenomenon as well—Peace is different in that the winner is not always associated with some manner of academic or research institution, or other part of research or industrial science. The award for Peace has gone to pacifists, civil rights activists, religious figures, and politicians/heads of state.

The prize for literature is the second award area where the United States does not have dominance. Here, both France and the United Kingdom surpass the United States' count and Switzerland matches it. All told, affiliation countries ring in at 35 and birth countries at 39.

In Economic Sciences, the dominance flexed by the U.S. borders on obscene. A staggering 76% of the awards have been won by someone affiliated with the United States. Affiliation countries ring in at a mere ten and birth countries at 19. The smaller numbers of overall countries are no surprise since this award has existed for roughly half the time of the other awards.

Another trend is observed if one considers the prize statistics more carefully, and this trend too is observed across multiple disciplines as well, with exceptions being Literature and Peace. Most pronounced in chemistry, earlier in the history of the Nobel Prize, it was less common, though not unheard of, for there to be more than one awardee in any given year. The award going to more than one person is increasingly common. Whether this reflects an

---

ECONOMIC SCIENCES

By affiliation

U.S.—68	U.K.—7	France—3	Germany—3	Norway—2
Sweden—2	Denmark—1	Finland—1	Netherlands—1	USSR—1

By birth

U.S.—56	France—4	U.K.—4	Canada—3	Netherlands—3
Norway—3	India—2	Russia—2	Russian Empire—2	Sweden—2
Austria—1	British Mandate of Palestine—1		British West Indies—1	
Cyprus—1	Germany—1	Hungary—1	Israel—1	Palestine—1
Scotland—1				

attempt to acknowledge more of the contributors, indecision, or a nod of sorts to the increasingly blended nature of sciences is difficult to pin down at this time, but it is a curious observation, nonetheless. Another possible explanation is that there is simply more science being done and with that increase in activity, there is a logical increase in the number of awards.

When considering a compilation of all the awards, a total of 64 countries have had someone affiliated win the Nobel Prize with birth countries totaling 91 different countries. Currently, the United States has a healthy and likely uncatchable lead in both the affiliation number and birth number. It is worth noting, however, that there is a disparity of just over 100 (113 to be exact) in the affiliation number versus birth number for U.S. awardees with birth trailing. This difference amounts to just over 25% of the U.S. awardees. This means that a healthy proportion of U.S.-based winners are immigrants of some kind.

What to do?

As it is virtually always the case that other work is at last as worthy as the awarded work, it is inevitable that this snubbed work was done by one or more women or someone with non-western origins. Without question, part of the problem is that these potentially deserving scientists are not even nominated. Since no contemporary list of nominations is published, it is impossible to know unless a committee member inappropriately leaks information or a nominator decides to disclose their nomination. Although causes for this potential lack of nomination are likely to be many, it is nearly impossible to exclude sexism or racism as one of them. We in the sciences may like to believe that we are above prejudices like racism, sexism and all the other "isms" in our society, focusing our attention instead on independent interpretations of data; instead, we, in fact, are not immune. Personal conflicts also inevitably rear their ugly heads. There are multiple pathways that may be effective at correcting this. For example, opening the nomination process more broadly or at least disclosing nominees may be low-hanging fruit for a more equitable process, assuming the cause is a lack of nominations for otherwise worthy awardees. This is because it would bring about a small measure of accountability or at least better transparency. As an academic, I like rubrics, a lot. A well-designed rubric that generates nominations for the committee may better recognize the work of these snubbed scientists. Even the award process could be made more fair, consistent, and equitable using a rubric. The internet and data analysis allow for all sorts of metrics to be accessed and assessed. Metrics include the number of publications; the number of citations, including citations per publication; and so many others can

ALL PRIZES

By affiliation

U.S.—404	U.K.—120	Germany – 83	France – 61	Switzerland—39
Sweden—31	Organization—27	Japan—19	Italy – 14	USSR—14
Denmark—13	Canada—12	Netherlands—11	Austria—10	Belgium—9
Ireland—7	Israel—7	South Africa—6	Spain—6	Australia—5
India—5	Poland—5	Argentina—4	Finland—4	Russia—4
China—3	Egypt—3	Chile—2	Czechoslovakia—2	East Timor—2
Greece—2	Guatemala—2	Hungary—2	Liberia—2	Mexico—2
Portugal—2	Bangladesh—1	Belgium and U.S.—1	Burma—1	Colombia—1
Costa Rica—1	Democratic Republic of Congo—1		Dutch East Indies—1	
Ethiopia—1	Germany and U.S.—1	Ghana—1	Iran – 1	Iraq—1
Kenya—1	Nigeria—1	Palestine—1	Peru and Spain—1	Philippines—1
Poland and U.S.—1		Prussia—1	South Korea—1	St. Lucia—1
Switzerland and U.S.—1		Tunisia—1	Turkey—1	U.S. and Japan—1
Vietnam—1	Yemen—1	Yugoslavia—1		

By birth

U.S.—291	U.K.—91	Germany—82	France—55	Sweden—28
Japan—28	Switzerland—20	Canada—20	The Netherlands—20	Italy—17
Russia—17	Russian Empire—17	China—11	Austria—14	Austria-Hungary—13
Denmark – 12	Norway– 12		Prussia—11	South Africa—10
Australia—10	India—10	Poland—10	Scotland—10	Belgium—9

(Continued)

(Continued)

ALL PRIZES

Ireland—8	Spain—8	Hungary—7	USSR —6	Egypt—6
West Germany—5	Argentina—4	Palestine—3	Austrian Empire—3	East Timor—2
British Mandate of Palestine—3	Mexico—2	Romania—3	Chile—2	Turkey—2
Liberia—2	Luxembourg—2	Portugal—2	Colombia—2	Schleswig—2
Iceland—2	Burma—1	New Zealand—2	French Algeria—2	Czechoslovakia—1
Ottoman Empire—2	Kenya—1	Israel—1	Finland—1	Iran—1
Guatemala—1	Vietnam—1	Costa Rica—1	Ethiopia—1	South Korea—1
Iraq—1	Mecklenburg—1	Nigeria—1	Philippines—1	Korea—1
St. Lucia—1	Morocco—1	Yemen—1	Bavaria—1	Württemberg—1
Lebanon—1	Tibet—1	Taiwan—1	Venezuela—1	Gold Coast—1
Hesse-Kassel—1	Trinidad and Tobago – 1	British India—1	Ethiopian Congo—1	Crete
Pakistan—1		Bosnia—1	Bulgaria—1	Peru—1
Guadeloupe Island—1		Madagascar—1	Persia—1	Ukraine—1
Tanzania—1		Cyprus – 1	Tuscany—1	
British West Indies—1				

be found by anyone literate with an internet connection. Such an approach would be defensible, consistent, and blind to every attribute of the researcher except for the work itself. Although it is possible that such a rigid evaluation method is already practiced, the small number of women awardees casts doubt, serious doubt in my opinion, on this possibility. Another flaw that can be argued is that work that is infamous or debunked may at times get enough "attention" by way of *negative* citations, that the rubric identifies it as worthy. This is where a human element would be able to (and ought to) override the rubrics. That said, opposition to such a rigid approach is not unfair. It can certainly be argued that such a rubric will disqualify a "dark horse" awardee. I at once disagree with this and find it irrelevant. First, a well-designed rubric can easily allow for such a nomination and award. Second, the Nobel Prize is not the NCAA Basketball Tournament, March Madness™. Nor is it the playoffs for some professional sport or even the Olympics where "Cinderella Stories" enthrall (and disappoint) millions. Although sometimes it takes years, decades even, to fully recognize and appreciate the enormous impact and importance, the Nobel Prize is not the place for a *feel-good underdog story*. It is for this reason that I believe a rubric that in some way evaluates the *already realized* impact of the works being nominated and considered is the fairest way forward. Quotas are one option that has been dismissed.[12]

The Chemistry Prize

Since its inception, through 2021, the chemistry award has been made 113 times. In 1916, 1917, 1919, 1924, 1933, 1940, 1941, and 1942, no award was given. One person, Frederik Sanger, was awarded the Chemistry Prize twice, though Linus Pauling and Marie Curie are chemistry laureates who have won two awards in a different field. Both of Pauling's awards (Chemistry and Peace) were unshared, and he remains the only *individual* to win more than one **unshared** Nobel Prize of any kind. Curie is one of two women to have earned an unshared chemistry award, the other being Dorothy Crowfoot Hodgkin in 1964 for her work on solving the structure of important biological substances using X-rays. Meanwhile, 60 men have taken home the prize solo. Once again, it is hard to justify this sort of disparity, even keeping in mind the growing number of instances where the award is given to more than one person. Curie remains the only woman to win more than one Nobel; one for physics (1903) and one for chemistry (1911).

The chemistry medal is the handiwork of Swedish sculptor and engraver Erik Lindberg. It represents nature in the form of a goddess resembling Isis. In her arms, she holds a cornucopia and is emerging from the clouds, a veil, covering her face, is held up by the Genius of Science.

FUTURE OF THE NOBEL PRIZE

There are no indications the Nobel Prize will cease to exist any time soon. Increasingly, some of the science-based awards are taking on more of a lifetime achievement award feel than one awarded for breakthroughs made in the preceding year. This does not take away from the prize. I cannot help but worry though that the especially Peace Prize will slowly but surely be increasingly accused of being political as some people are awarded it while others denied it.

Increasingly controversial is who receives recognition for the work done vis-à-vis the Nobel Prize, and it is here I think some changes may come eventually. Although the Literature Prize and, perhaps, the Peace Prize are justifiably a truly individual award, scientific endeavors are more collaborative today than in Nobel's time. I would argue, in fact, that Nobel and his contemporaries could not have foreseen just how collaborative science would become. Not only are projects often conducted by teams of several researchers, rather than an individual, but the nature of any individual project is also increasingly collaborative since modern science is increasingly blended and interdisciplinary. As a result, it often demands multiple fields of expertise. This inevitably leads to a larger number of people. Projects also now take far longer the complete than 120 years ago and the increase in time only further increases the number of people involved. This brings about a natural, and exceedingly difficult to answer the question of who among all these collaborators deserve(s) the award. Often, the PI, the Principal Investigator—the boss for the reader unfamiliar with the terminology—is the one who receives the award. This, even though in modern science this individual is very unlikely to have performed even one of the experiments behind the work. Why then do they get the award? It is as complicated as it is unfair.

First and foremost, this individual is *the constant*; they are the person who has been involved in the project from the beginning and is the only person (usually, anyway) who has had input to add to every aspect of the project since its inception. More for the inexperienced reader than the experienced reader, this individual is also the person who pays most of the (scientific) bills of the research through their grants. They also, whether they are in the lab doing actual scientific work or not contribute heavily to the projects by way of suggestions for future/additional experiments and problem-solving and in interpreting results. That said, there is some validity to complaints that such an approach is unfair and that it fails to appropriately acknowledge the work of the people performing the experiments in the laboratory. Unfortunately, I do not see any viable alternatives currently though it is for sure something to attempt to right. There has even been a suggestion that the Nobel Prize should be given for a topic, rather than to people.[13] How this relates to decidedly individual prizes such as Peace and Literature is unclear to me.

Future of the Chemistry Nobel Prize

A survey of what specifically the Nobel Prize in Chemistry is awarded for shows a clear rise of biochemistry. Anecdotally, a colleague of mine has lamented that the Chemistry Nobel Prize seems to be more biology sometimes than chemistry to them. The Physiology or Medicine Prize is also often as much biochemistry as it is physiology or medicine. Time will tell if the focus of the chemistry award continues or shifts to more energy-based science.

YEAR	SUBDISCIPLINE[a]	YEAR	SUBDISCIPLINE[a]	YEAR	SUBDISCIPLINE[a]	YEAR	SUBDISCIPLINE[a]	YEAR	SUBDISCIPLINE[a]
1901	P	02	O	03	G	04	G	05	O
06	G	07	B	08	N	09	G	10	O
11	N	12	O, O	13	I	14	E	15	B
16	None	17	None	18	I	19	None	20	P
21	N	22	A	23	A	24	None	25	G
26	G	27	B	28	B	29	B, B	30	B
31	G, G	32	I	33	None	34	N	35	N, N
36	A	37	B, B	38	B	39	B, O	40	None
41	None	42	None	43	P	44	N	45	B
46	B, B, B	47	O	48	A	49	P	50	O, O
51	N, N	52	A, A	53	I	54	G	55	B
56	P, P	57	B	58	B	59	A	60	EV
61	EV	62	B, B	63	M, M	64	A	65	O
66	G	67	A, A, A	68	P	69	G, G	70	B
71	G	72	B, B, B	73	O, O	74	P	75	O, O
76	I	77	P	78	B	79	O, O	80	B, B, B
81	O, O	82	A	83	I	84	O	85	A, A
86	P, P, P	87	O, O, O	88	P, P, P	89	B, B	90	O

(Continued)

(Continued)

YEAR	SUBDISCIPLINE[a]	YEAR	SUBDISCIPLINE[a]	YEAR	SUBDISCIPLINE[a]	YEAR	SUBDISCIPLINE[a]	YEAR	SUBDISCIPLINE[a]
91	A	92	P	93	B, B	94	O	95	EV, EV, EV
96	O, O, O	97	B, B, B	98	P, P, P	99	A	2000	M, M, M
01	O, O, O	02	A, A, A	03	B, B	04	B, B	05	O, O, O
06	B	07	M	08	B, B, B	09	B, B, B	10	O, O, O
11	I	12	B, B	13	G, G, G	14	A, A, A	15	B, B, B
16	G, G, G	17	B, B, B	18	B, B, B	19	EN, EN, EN	20	B, B
21	O, O								

[a] *Subdiscipline key:* A, analytical chemistry/instrumentation; B, biochemistry; EG, energy storage; EV, environmental; G, general chemistry; I, inorganic chemistry; M, materials; N, nuclear; O, organic chemistry; P, physical chemistry.

Predictions

If for no other reason but to have a little fun, I would like to make a few predictions on future Nobel Prizes. For soothsaying, I am going to restrict my actual predictions to the Physics, Chemistry, and Physiology or Medicine awards, though I will comment on the Peace Prize briefly as well. Regarding the chemistry Nobel Prize, as the synthetic protocols currently referred to as C–H activation see further development, it is highly likely to earn someone a Nobel Prize. It is currently too early to identify a true frontrunner, but Melanie Sanford and M. Christina White are two leaders in the field currently. Alternatively, carbon sequestration or the conversion of carbon dioxide back to gasoline or other oil products are also important endeavors that would be the carbon and environmental equivalent to Haber's Nobel-winning work. Although it is without question worth asking "why would we remake oil products from carbon dioxide?", it is important to recognize that the oil industry provides way more than just energy. It also provides starting materials and solvents for various critical materials such as textiles, dyes, and arguably most important of all, the pharmaceutical industry. An alternative source of these important substances or fuel will be very important as these non-renewable resources dwindle in their supply and/or access due to geopolitical conflicts and increase in their cost.

Taking physics next, if life—even simple life—were ever to be discovered elsewhere in the solar system (e.g., Mars), those responsible for doing the experiments or at least those who designed them will almost certainly win a Nobel Prize. Such a discovery would have incomprehensible ramifications for humankind. If recreational space travel becomes more popular, it is plausible (in my opinion, anyway) that either Elon Musk (*via* Space X) or Jeff Bezos (*via* Blue Origin) may be in line for a physics Nobel Prize. In fact, what those two have done already may be worthy of a Nobel Prize. The development of the technologies by an independent company rather than a governmental entity is utterly amazing. The development of the technologies each company uses is as sophisticated or more as anything any Nobel Prize has been previously awarded for. The automated and reusable rockets are technological wonders to be sure. The other thing they have potentially done is popularized something related to science. As of this writing, recreational space travel accessible to the masses is a long way off. However, the first steps have been taken because of these efforts. All that said, going to the moon did not earn anyone a Nobel Prize so it is perhaps foolhardy to think this will.

Technology related to renewable energy also may eventually earn someone a Nobel Prize in Physics. However, improvements that would likely be needed to make that a reality would need to be significant, either in the reduction of cost, increase of yield/conversion, or both. Actually, depending on the precise nature of the technology, this sort of breakthrough may, in fact, be more worthy of a Chemistry Nobel Prize than a Physics one. It is shocking to me that energy storage has only been part of one Nobel Prize, the prize in 2019 for lithium-ion batteries, awarded to John B. Goodenough, M. Stanley Whittingham, and Akira Yoshino.

Both the electric and self-driving represent technological breakthroughs absolutely worthy of a Nobel Prize, and Elon Musk also has a hand in. Both have enormous advantages over current technologies and as each improves, we are increasingly likely to reap all those benefits and more. It is difficult though to place this sort of breakthrough into one of the categories covered by the Nobel Prize.

For Physiology or Medicine, it is low-hanging fruit to predict that the m-RNA technology that led to the most effective of the COVID-19 vaccines will earn someone the Nobel Prize. This will be especially true if it is found that this technology is universally applicable to other vaccines as well and I predict it will be. Cancer vaccines and even anticancer manmade viruses have been developed and if they prove successful would without question be at or near the front of the line for the award.

Finally, although not exactly a prediction, I must admit I am shocked that the Nobel Peace Prize has not yet been awarded to the Bill and Melinda Gates Foundation. The philanthropy done by this organization to date is tremendous and world-changing. Because of this generous work, hundreds of millions of people in some of the poorest and most disease-ridden parts of the world have access to medicine. The foundation has also done amazing work to improve education and access to technology in classrooms, globally. My only guess as to why no Nobel Prize has been awarded to the foundation is an attempt to send a message that you cannot "buy" a Nobel Peace Prize. Nothing other than that sort of message can explain to me why this foundation has not been bestowed this (in my opinion deserving) honor.

The prize and society

Shortly after the announcement of the 2021 Physics Prize, an opinion appeared on CNN claiming, "This Nobel Prize is a game-changer."[14] This reflects the opinion of the article's author that—since the Nobel Prize here is related to climate change—this should turn the tides of belief in favor of the validity of anthropogenic climate change. This is because of an implied vouching (my words) done by the committee for the science by making this award. In my opinion, this is untrue. Those denying anthropogenic climate change are already willfully ignoring the widely agreed upon science. One more group of scientists "endorsing" it is very unlikely to perturb their stance. Additionally, there are many people who (apparently without jest) believe the world is flat and/or that the world is a mere several thousand years old. Both beliefs are held even when confronted with mountains (pun very much intended) of scientific evidence. In fact, I may argue the exact opposite stance about the societal impacts of the award. The deniers may, in fact, lose even more faith in the scientific establishment, viewing the award as something agenda-driven, rather than science-driven. Such possibilities—turning the tides of belief, further entrenching against, or anything in between—should **never** be the goal of the award. The goal of the Nobel Prize never was, is not, and never will be to convince the throes of non-science persons to believe a scientific finding.

ABOUT THIS SERIES OF BOOKS

This series of books focuses on the Nobel Prizes in Chemistry that have contributed to the field of synthetic organic chemistry. Such a broad scope necessitates decisions that may appear arbitrary to the reader. I assure you, no slight is meant by any decisions to exclude certain awards, especially those in the next section. For certain, awards other than those covered in these volumes *involve* organic chemistry, even synthetic organic chemistry. Herein, I have chosen to omit any that only utilize or apply synthetic organic chemistry, rather than build it up. A section covering "honorable mention" Nobels attempts to justify some of the omissions. In essence, I have tried to create a story of how the Nobel Prize in Chemistry made synthetic organic chemistry what it is today. Are such lines blurry? In a word, yes, and to be fair, even I think some of the awards I have chosen to include are borderline, at best. As the field continues to develop, I yield that some currently omitted studies may merit inclusion. If such a scenario comes about, I will address it with later editions, volumes, and/or both.

Currently, the volumes are delineated chronologically. Such an organization was chosen to allow for the easier creation of both later, updated editions and future volumes. Furthermore, a thematic organization would be difficult, if not impossible to balance and adhere to parameters of logical and comparable volume sizes. Additionally, in some years when the prize is awarded for multiple works, one may be more related to organic synthesis than the other. In this sort of case, the unrelated work will be mentioned but not receive a detailed review. The mention will be restricted to the bare minimum necessary to retain historical accuracy for the award.

Volume 1

1902 Fischer "in recognition of the extraordinary services he has rendered by his work on sugar and purine syntheses"

1910 Wallach "in recognition of his services to organic chemistry and the chemical industry by his pioneer work in the field of alicyclic compounds"

1912 Grignard "For the discovery of the so-called Grignard reagent, which in recent years has greatly advanced the progress of organic chemistry"

and Sabatier "for his method of hydrogenating organic compounds in the presence of finely disintegrated metals whereby the progress of organic chemistry has been greatly advanced in recent years"

1950 Diels and Alder "for their discovery and development of the diene synthesis"

1965 Woodward "For his outstanding achievements in the art of organic synthesis"

Volume 2

1979 Brown and Wittig "for their development of the use of boron (Brown) and phosphorus (Wittig)-containing compounds, respectively, into important reagents in organic synthesis"

1981 Fukui and Hoffmann "for their theories, developed independently, concerning the course of chemical reactions"

1990 Corey "for his development of the theory and methodology of organic synthesis"

Volume 3

1994 Olah "for his contribution to carbocation chemistry"

2001 Knowles and Noyori "for their development of catalytic asymmetric synthesis" and "for their work on chirally catalyzed hydrogenation reactions"

and Sharpless "for his work on chirally catalyzed oxidation reactions"

2005 Chauvin, Grubbs, and Schrock "for development of the metathesis method in organic synthesis"

Volume 4

2010 Heck, Negishi, and Suzuki "for palladium-catalyzed cross couplings in organic synthesis"

2018 Frances H. Arnold "for the directed evolution of enzymes" (shared with **Smith and Winter** "for the phage display of peptides and antibodies," both of whom are not covered here)

2021 List, MacMillan "for the development of asymmetric organocatalysis"

Honorable mentions

As one might imagine, choosing who to include and more importantly not include in this sort of compilation is a challenging task. I dare (jokingly) quip that this decision is even more difficult than awarding a Nobel in the first place as I am assuming the position of omitting recognized achievement. Perhaps, only to relieve a guilty conscience of sorts, I mention—and attempt to justify—here a handful or so of my most difficult decisions to omit in chronological order.

1905—Adolf von Baeyer "in recognition of his services to the advancement of organic chemistry and the chemical industry, through his work on organic dyes and hydroaromatic compounds." Arguably, this omission is the most egregious. His contributions to the field of organic chemistry are without doubt great and far-reaching. However, as far as their improving or building up the field of synthetic organic chemistry, not only is his primary contribution to this field (the Baeyer-Villiger reaction) not related to his Nobel, his award-winning work does not offer new and versatile synthetic methods or a changing of the way synthetic chemistry is conceived and performed. Perhaps, future editions will include Baeyer, for now, I stand by my decision to omit it.

1923—Fritz Pregl "for his invention of the method of microanalysis of organic substances" (for chemical composition). The importance of microanalysis of organic substances cannot be overstated. This work allowed for nothing less than the far easier determination of the molecular formula of chemical substances. Although running the experiment may not be, performing the calculations using data from this experiment are a standard part of many, if not all, college-level general chemistry classes. Without such data, identifying the chemical structure of compounds would be far more arduous. All that taken into consideration, this does not contribute to synthetic organic chemistry, so it is omitted here.

1947—Sir Robert Robinson "for his investigations on plant products of biological importance, especially the alkaloids." This work showed troponin alkaloids can be made from three simpler molecules. Omitting Robinson may even be more egregious than omitting Bayer. What Robison showed is (effectively) complicated molecules (at least the tropin alkaloids) could be made from much simpler building blocks. Had R. B. Woodward and later Corey not *totally* changed and expanded what was possible, it would be harder to omit Robinson. The outstanding work by Woodward and Corey, however, sets the bar so much higher that it is too difficult to include Robinson.

1956—Hinshelwood and Semenov "for their researches into the mechanism of chemical reactions." It is inarguable that a better understanding of chemical reaction mechanisms—how on an atomic level and/or molecular level molecules rearrange to go from starting materials to products for the less experienced reader—helped to drive the field of organic synthesis forward. However, as it is difficult to impossible to identify a specific way that this work bult synthetic organic chemistry, it is omitted here.

1984—Robert Bruce Merrifield "for his development of methodology for chemical synthesis on a solid matrix." Merrifield's award is basically for polymer-bound peptide synthesis and without question has totally changed peptide synthesis. Without question, his contribution has revolutionized synthesis. However, to date, this has been primarily focused on peptide synthesis quite narrowly. If it were to ever be expanded to more general synthetic organic methods, it would be harder to justify the omission. Some work has been done to apply it to general organic synthesis, but in my opinion, more must be done for a work such as this collection.

2018—Smith and Winter "for the phage display of peptides and antibodies" while including Arnold "for the directed evolution of enzymes." In short, Arnold's work has potential to lead to enzymes being developed that will permit easier chemical transformations, perhaps—in a best-case scenario—leading to reactions that can compete with the selectivities and high yields observed with biological systems *in vivo*. Meanwhile, the work of Smith and Winter has no relation to synthetic chemistry.

GENERAL REFERENCES

https://www.nobelprize.org/, last checked 6/29/22.
https://chemistry.as.miami.edu/_assets/pdf/murthy-group/gnl_jensen-2.pdf, last checked 6/29/22.

NOTES

1 https://www.nobelprize.org/nomination/chemistry/, last checked 6/29/22.
2 https://www.nobelprize.org/nomination/physics/, last checked 6/29/22.
3 https://www.nobelprize.org/nomination/medicine/, last checked 6/29/22.
4 https://www.nobelprize.org/nomination/literature/, last checked 6/29/22.
5 https://www.nobelprize.org/nomination/peace/, last checked 6/29/22.
6 https://www.nobelprize.org/nomination/economic-sciences/, last checked 6/29/22.
7 https://www.chemistryworld.com/features/are-the-nobel-prizes-good-for-science/3009557.article, last checked 6/30/22.
8 https://www.usnews.com/news/best-countries/articles/2020-10-01/the-nobel-prizes-have-a-diversity-problem-worse-than-the-scientific-fields-they-honor, last checked 6/30/22.
9 https://www.science.org/content/article/one-reason-men-often-sweep-nobels-few-women-nominees, last checked 6/30/22.
10 Borman, Stu; Chemical and Engineering News, September 11th, 2017, 22–24. "Women Overlooked for Nobel Honors."
11 https://www.popsci.com/racial-inequality-nobel-prize/, last checked 6/29/22.
12 https://www.bbc.com/news/world-europe-58875152, last checked 6/30/22.
13 https://massivesci.com/articles/nobel-prize-science-gender-physics/, last checked 6/30/22.
14 https://www.cnn.com/2021/10/06/opinions/physics-nobel-climate-change-lincoln/index.html, last checked 6/29/22.

1979—Brown and Wittig

2

The 1979 Nobel Prize in Chemistry was awarded to two people—Herbert C. Brown and Georg Wittig—for two different discoveries related to synthetic organic chemistry. Like Grignard and Sabatier in 1912, the pair shared the Nobel Prize for two quite different chemical reaction series. Brown's portion of the award was for the development and use of boron-containing compounds, and Wittig's was for phosphorus-containing compounds. Although of the two, only Wittig (pronounced Vittig, by the way) has a chemical reaction that bears his name, the scope of the chemistry Brown contributed to is truly huge. Both Wittig and Brown invented chemical transformations that are used—even if with modest modifications in some cases—daily in the synthesis of a myriad of products including compounds of pharmaceutical interest, commercial interest, and, of course, complex natural product synthesis.

Starting with Brown, before delving into his beautiful chemistry, a beautiful human story must be (re)told, a story he gives an account of in his Nobel lecture. In the lecture, Brown tells the story of how he first became interested in boron hydrides. As he tells it, his girlfriend and future wife bought him a graduation present of a book on this topic after his undergraduate studies were completed. She did not select the book because Brown had expressed an interest in the topic during some conversation and she remembered it. She bought it because it was the "most economical chemistry book available at the University of Chicago bookstore." The price ($2.06) is shocking to all of us today. According to an online inflation calculator,[1] this is equal to $43.91 today. Even this price seems cheap, though if the book is smaller, it may be consistent with today's prices. Stories like this impart a human element, in my opinion, that allows any of us to have a personal connection to the award winners.

Inspired by the chemistry contained in the book, Brown pursued this in his Ph.D. studies. Shortly before Brown started his graduate work, it was found that carbon monoxide and diborane reacted to form boron carbonyl H_3BCO. The actual nature of the product was not well understood at the time (modern spectroscopy was still years away). To try to solve this problem,

DOI: 10.1201/9781003006848-2

Brown was encouraged to explore diborane's reactivity with aldehydes and ketones. Once Brown mastered the sensitive and complicated laboratory techniques, he got to work on a series of aldehydes, ketones, esters, and acid chlorides. He found that all but the acid chlorides resulted in the initial formation of dialkoxyl boranes that rapidly hydrolyze to the corresponding alcohols in water. Brown then notes that at the time of this work, not only were there no genuinely effective ways to reduce a carbonyl, not much was being done about this synthetic shortcoming. Brown's culprit in this is the rarity of diborane. The solution came from perhaps as unlikely a source as Brown's inspiration: World War II, specifically, nuclear research even if only indirectly. Brown, while working in the Schlesinger lab at USC investigated the use of many metal borohydrides in pursuit of low-molecular-weight uranium compounds. The complete story of uranium would be a distraction here, so it is omitted, but the interested reader is encouraged to read Brown's Nobel lecture and other sources for more information on this topic.

Upon taking a position at Purdue University in 1947, Brown began the synthetic organic work that is the cornerstone for hydride reductions. At this point, sodium borohydride ($NaBH_4$) and lithium aluminum hydride ($LiAlH_4$) were the two known reducing agents. The former's reductivity was discovered by Brown while attempting to design his efficient synthesis of it. It turns out that these two cases effectively represent the two extremes of hydride reduction power, though this is admittedly an oversimplification on my part. Lithium aluminum hydride is a powerful reducing agent and covers a wide range of functional groups that will submit to it. Brown and his group set out to identify reagents that would give synthetic chemists at least some measure of selectivity. At the time of the work, any selectivity at all would be an improvement. Two trends were found very quickly with one being based on the metal—$LiBH_4 < Mg(BH_4)_2 < Al(BH_3)_3$; and the other based on the presence of alkoxy substituents—$LiAlH(O\text{-}t\text{-Bu})_3 < LiAlH(OMe)_3 < LiAlH_4$. They even found that limits could be pushed in examples like $K(i\text{-}PrO)_3BH$ being less reactive than $NaBH_4$ and $LiEt_3BH$ being more reactive than $LiAlH_4$. Ultimately, the work led to a wide range of reducing agents (and others have been developed since) that could be used in a variety of settings. Table 2.1 summarizes the results obtained by Brown and his coworkers. In the table, a+ is used to indicate that a reduction occurs; a− is used to indicate a reduction does not occur and a± is used to indicate variable reactivity, which is sometimes the reduction proceeds, sometimes it does not.

In all these reactions, a hydride is delivered to the molecule, specifically to the carbon atom, which carries a greater partial positive charge while the oxygen carries a partial negative charge. During the workup of the chemical reaction, the alkoxyl intermediate is hydrolyzed, delivering a proton to the oxide oxygen or the amide nitrogen.

TABLE 2.1 Summary of reduction conditions. A = $NaBH_4$/EtOH, B = $Li(t\text{-buO})_3AlH$/THF, C = $LiBH_4$/THF, D = $Al(BH_4)_3$/diglyme, E = B_2H_6/THF, F = sia_2BH/THF, G = 9-BBN/THF, H = AlH_3/THF, I = $Li(OMe)_3$/THF, J = $LiAlH_4$/THF

FUNCTIONAL GROUP	A	B	C	D	E	F	G	H	I	J
Aldehyde	+	+	+	+	+	+	+	+	+	+
Ketone	+	+	+	+	+	+	+	+	+	+
Acid chloride	+[15]	+	+	+	−	−	+	+	+	+
Lactone	−	±	+	+	+	+	+	+	+	+
Epoxide	−	±	+	+	+	±	±	+	+	+
Ester	−	±	+	+	±	−	±	+	+	+
Acid	−	−	−	+	+	−	±	+	+	+
Acid salt	−	−	−	−	−	−	−	+	+	+
Tertiary amide	−	−	−	−	+	+	+	+	+	+
Nitrile	−	−	−	−	+	−	±	+	+	+
Nitro	−	−	−	−	−	−	−	−	+	+
Olefin	−	−	−	−	+	+	+	−	−	−

This closes the book on the reduction series that Brown elegantly developed. But the story of Brown's Nobel does not stop here by any means. An offshoot, hydroboration, is every bit as important as the polar bond reductions. As Brown tells the story, this was discovered somewhat on accident when an anomaly was uncovered in the number of equivalents of reagent consumed during the reduction of ethyl oleate. In addition to the ester moiety, ethyl oleate contains a lone olefin (a carbon-carbon double bond, an alkene). In this case, almost 20% more reducing agent was consumed. Closer investigations revealed that not only was the carbonyl (the carbon–oxygen double bond) reacting, but the alkene was also reacting such that a hydrogen atom is added to one atom of the (now former) alkene and a boron to the other, reducing the alkene to an alkane; specifically, an organoborane.

A brief sidebar is opportune here. Recognize that a portion—a sizable portion in fact—of Brown's borane work arose from an anomaly. This, in my opinion, is equal parts beautiful science and stellar flexibility. First, ascertaining what was occurring, rather than what was expected is simply good science. That its implications could also be followed to this degree is exemplary. The perseverance needed to pursue a different direction in the face of the anomaly is also praiseworthy. Brown also laments the early naysayers claiming the organoboranes were not a worthy pursuit. I dare say he proved them wrong and may make such perseverance even more impressive.

Anyone who has published knows that reviewers can be fussy and that work that challenges paradigms or reports something *brand new* is often difficult to get accepted by the peer review process. The sentiment expressed by the naysayers would no doubt encourage a certain intellectual prejudice against this chemistry and yet Brown and his coworkers carried on.

During the optimization of this hydroboration reaction (Figure 2.1), it was found that ethers catalyze the reaction that converts alkene **1** into alkyl borane **2**. The dependence on ethers is convenient since ethers are quite common solvents for organic reactions on account of their general nonreactivity. Moreover, the reaction typically proceeds very rapidly and is high yielding. A postdoc in Brown's lab (Dr. Subba Rao) developed the procedure to convert the initially formed organoborane into the corresponding alcohol **7**. It was also established that the reaction proceeds to the anti-Markovnikov—the less substituted—product. Brown and coworkers further established the stereochemistry of the addition showing that the hydrogen and the boron are added cis (pointing in the same direction) with the eventual oxygen retaining the three-dimensional position of the boron **4–7**. In sterically hindered systems this addition occurs on the less sterically hindered face of the reacting alkene/olefin **8–9**. Finally, the addition and subsequent conversion to the alcohol never rearranged. This lacking of rearrangements yields valuable information about the mechanism of the reaction, specifically that it does not involve the formation of carbocations as intermediates.

In fact, both the *syn* addition of the H-B and the lack of rearrangements can be explained by the mechanism of the addition. According to the accepted mechanism, the hydrogen atom is delivered (effectively as hydride) while the boron atom accepts electrons from the pi system. This mechanism also explains the formation of the less substituted product. The boron, acting as a Lewis acid accepts electrons from the less substituted carbon atom, where a negative charge would be more stable. Meanwhile, the hydrogen atom is delivered as hydride (so with its electrons as a Lewis base) to the

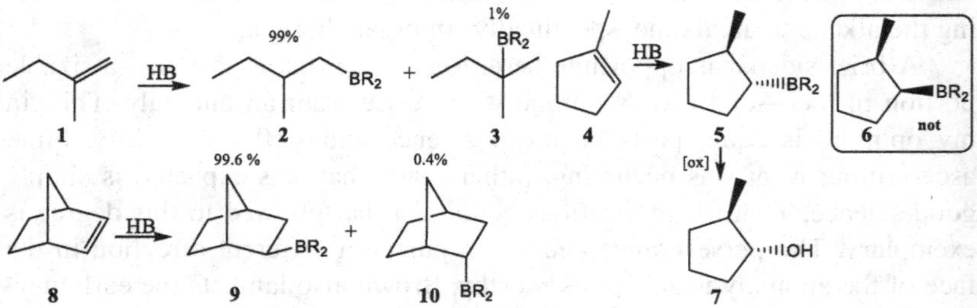

FIGURE 2.1 Model alkene hydroboration reactions.

more substituted carbon atom where a positive charge would be more stable. In short, although such charges are certainly *not formed*, the system behaves as if they are formed at least from the point of view of which atom reacts with which.

Shown in Figure 2.2 is a concerted addition of the hydrogen and boron to both alkene **11** to give alkyl borane **12** and alkyne **13** to give vinyl borane **14**. The result of such a concerted (one step) mechanistic pathway is that no charged intermediates are ever formed. In each of the cases considered here, if this were a stepwise mechanism whereby the boron is added first, the following cations shown in Figure 2.3 would have been generated. If an alkene reacts with BH_3 to form a cation, **15** would be generated and would rapidly rearrange to the more stable tertiary cation **16**. Meanwhile, the alkyne would give the vinyl cation **17**, which would be in equilibrium with the trans isomer **18**. Since the trans isomer is more stable, the equilibrium would favor its formation rather than the cis isomer.

In practice, neither the rearrangement shown in the case of the alkene nor the equilibrium shown in the case of the alkyne occurs, thus, the formation of these cations cannot be invoked; none of the outcomes of cation formation are observed. The concerted addition described earlier accounts for the lack of rearrangements and lack of equilibrium because it plausibly describes how the bonds are formed without invoking the formation of a cation. It also accounts for syn addition, rather than anti-addition to both the alkene and alkyne. This observation is best understood by invoking collision theory. Remember that for a reaction to occur, a molecular collision must take

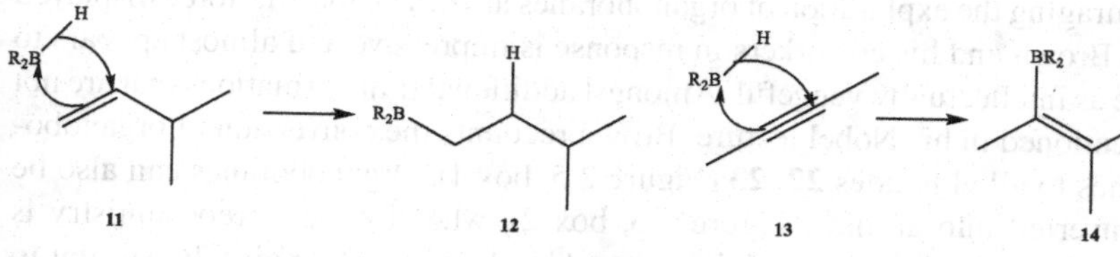

FIGURE 2.2 Alkene and alkyne hydroboration mechanisms.

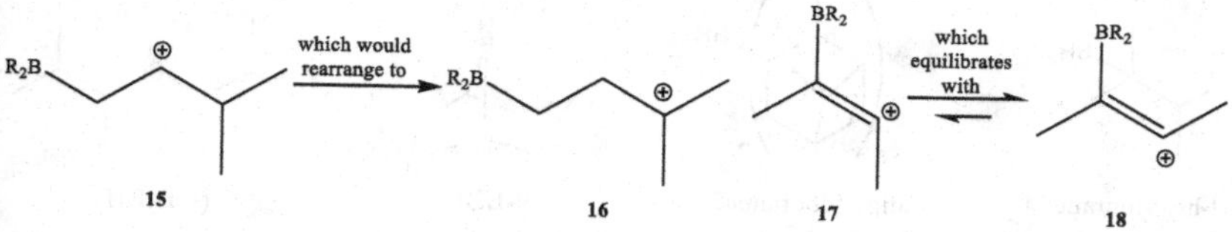

FIGURE 2.3 Intermediates of alkene and alkyne hydroboration assuming the formation of a carbocation.

place. If the face (side) of one alkene carbon is colliding with the boron atom, the same face of the other alkene carbon atom is more likely to simultaneously collide with a hydrogen atom attached to the same boron atom than to a hydrogen atom attached to a different boron atom (and, thus, molecule) on the other side (face).

Armed with this understanding, a variety of hydroborating agents **19a–d** then followed from Brown's lab. Usually, when simple borane (BH_3) is used, the boron becomes trialkylated by the alkene or alkyne. Mono- and dialkyl borane derivatives (Figure 2.4) have the notable advantage of being more sterically hindered, which helps to further drive the reaction towards the less substituted product. Meanwhile, dipinylborane **19b** allows for a not just regioselective but stereoselective reaction of internal akenes.

All of this is well and good, but there is an important aspect of this process at least when using alkyl boranes as reagents. This process suffers from what is called poor atom economy. Atom economy can essentially be summarized as the percent of atoms across all starting materials and reagents that are found in the desired product. Most often, these pre-existing alkyl groups attached to the boron do not find their way into the product. This arguably makes them wasteful. Moreover, in some applications (in the case of the oxidation step of this sequence, for example), by-products are also formed, necessitating their removal via purification. This issue of atom economy is not unique to hydroboration reactions. However interesting the topic of atom economy may be (and the related step and pot economies), a discussion past this point is beyond the scope of this book

Previously, it was mentioned that Brown lamented detractors earlier discouraging the exploration of organoboranes at all. The tour-de-force displayed by Brown and his coworkers in response is impressive and almost appears to me as intellectually vengeful. Amongst additional transformations that are not mentioned in his Nobel lecture, Brown recounts the conversion of organoboranes to alkyl halides **22, 23** (Figure 2.5, box 1). Organoboranes can also be converted into amines (Figure 2.5, box 2), whereby the stereochemistry is retained at the boron, which is replaced by the nitrogen to give **26**, similar to what is seen in the conversion to alcohols. Using a halohydroborane as the

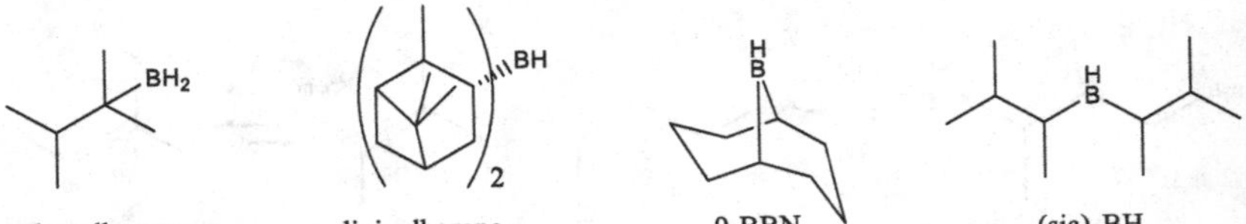

t-hexylborane	dipinylborane	9-BBN	(sia)$_2$BH
19a	**19b**	**19c**	**19d**

FIGURE 2.4 Sample hydroboration reagents.

boronylating agent, thereby generating an alkyl haloborane, the ordinarily sluggish reaction with alkyl azides proceeds nicely to secondary amines such as **29**. Arguably, some of the most important transformations these species are capable of are the carbon–carbon bond forming reactions (Figure 2.5, box 3), which the cyclopropanation is as well. In these other carbon–carbon bond forming reactions, the molecule is elongated, the importance of which to synthetic organic chemistry cannot be overstated. For example, the use of silver derivatives. Ketones, esters, and nitriles can also be alkylated or arylated at the α position (the position immediately neighboring the carbonyl) using alkyl boranes. All of this before even mentioning the Pd-cross coupling reactions that use alkyl borane reagents, in particular, the Suzuki coupling reaction. They can also be used to make cyclopropanes like **42** if the starting alkene is appropriately substituted (Figure 2.5, box 4).

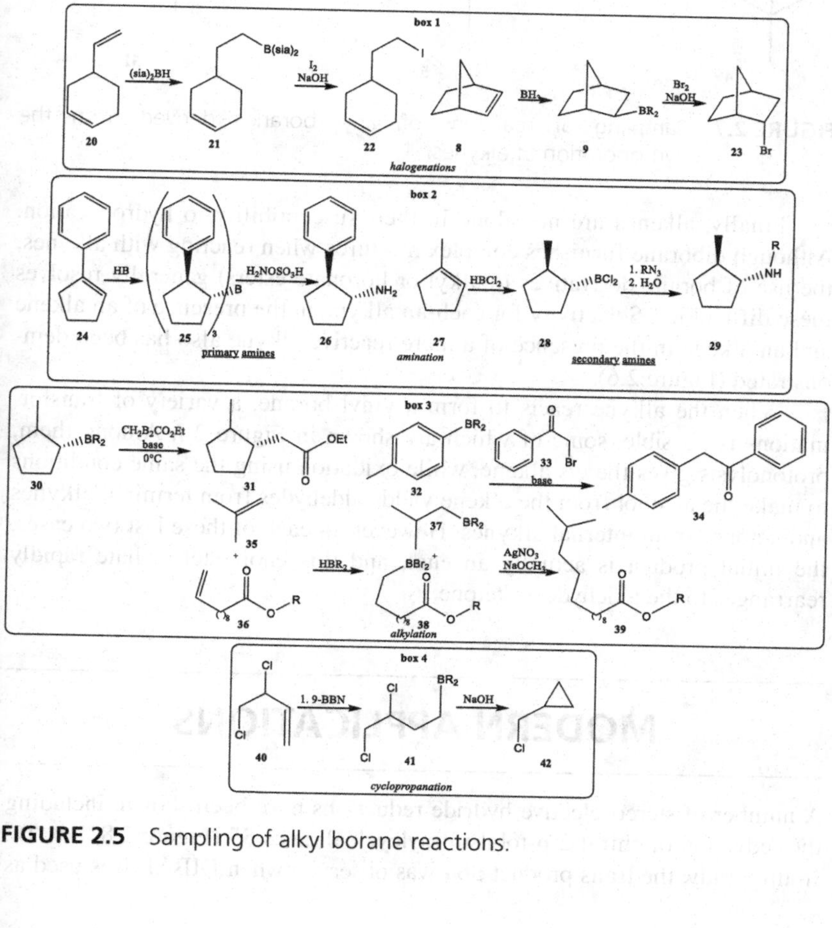

FIGURE 2.5 Sampling of alkyl borane reactions.

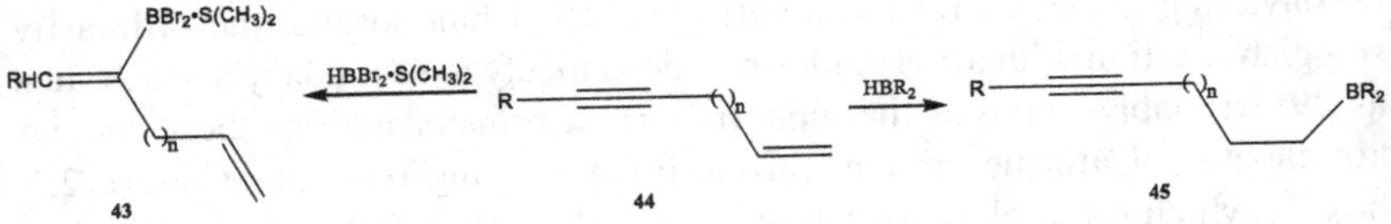

FIGURE 2.6　Selective hydroborations of alkenes and alkynes.

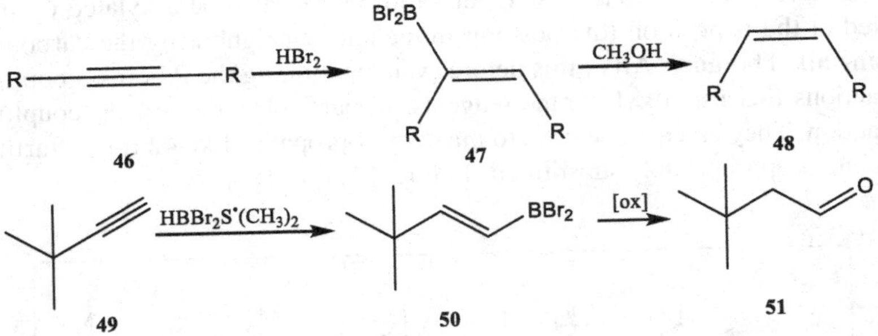

FIGURE 2.7　Sampling of reactions of vinyl boranes derived from the hydroboration of alkynes.

Finally, alkenes are not alone in their susceptibility to hydroboration. Although diborane furnishes complex mixtures when reacting with alkynes, the use of borane derivatives (dialkyl or boronate esters) generally resolves these difficulties. Selectivity for each-an alkyne in the presence of an alkene and an alkene in the presence of a more reactive alkyne also has been demonstrated (Figure 2.6).

When the alkyne reacts to form a vinyl borane, a variety of transformations is possible, some of which are shown in Figure 2.7. Among them, protonolysis gives the cis alkene, while oxidation using the same conditions to make the alcohol from the alkene yields aldehydes from terminal alkynes and ketones from internal alkynes. However, in each of these last two cases, the initial product is actually an enol, and this enol intermediate rapidly rearranges to the aldehyde or ketone.

MODERN APPLICATIONS

A number of stereoselective hydride reductions have been shown, including the reduction of chiral 2-p-tolylsulfinylcycloalkanes **55** (Figure 2.8, box 1).[2] In this study, the trans product **56a** was observed when DIBAL was used as the hydride-reducing agent and when DIBAL/ZnCl$_2$, LAH, NaBH$_4$, LiEt$_3$BH, and Li(s-Bu)$_3$BH are used, the cis product **56b** is observed as the major product. The overall yields consistently ranged from 60% to 91%. The reduction of imines[3] **57** has also been shown using HBpin and an iron catalyst, furnishing the amine **58** in high yields (Figure 2.8, box 2). This study showed that 5 mol % FeCl$_2$ and 10 mol % Mg along with HBpin, followed by treatment with silica or HCl smoothly furnishes the amine from the corresponding imine. A variety of aromatic rings were used including heteroaromatic compounds in the series described. Borane-catalyzed pyridine hydroboration-reliant trifluoromethylthiolation and difluoromethylthiolation have been shown[4] to be effective for late-stage functionalization of pyridine drugs for the generation of new drug candidates (Figure 2.9). Among others, this work made trifluoromethylthiolated analogs of etoricoxib **63**, which is used in the treatment of rheumatoid arthritis and other pain ailments, and etofibrate **64**, a cholesterol-lowering medicine. As of this writing, it does not appear that either analog has undergone pharmacological evaluation. Coleman and Gurrala[5] (Figure 2.8, box 4) used a diastereoselective hydroboration-oxidation in their synthesis of eupomatilones 4 and 6, correcting the structure of eupomatilone 6 along the way. At the time of their discovery,[6] these compounds had novel structures. Novel structures often attract synthetic interest. Impressively enantioselective hydroborations of silyl enol ethers have also been observed using a cobalt catalyst (Figure 2.8, box 3). In this work,[7] aryl ketone-derived silyl enol ethers **59** were in high yield and enantioselectivity hydroborated and then oxidized. A number of pharmacologically important compounds or natural products have hydroxyl or alkoxyl groups in the benzylic position (Figure 2.10) and can potentially be prepared more efficiently via a route such as the one described here. It is worth noting that even in cases where the stereoselectivity of the reaction furnishes the isomer where the hydroxyl group is in the opposite direction than the one desired, its stereochemistry can be inverted easily using a Mitsunobu reaction.[8]

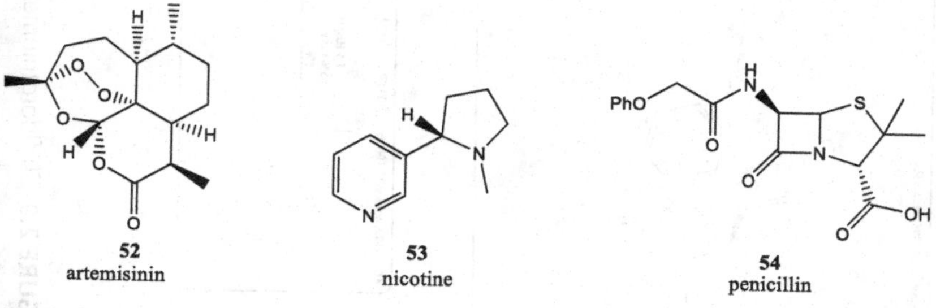

52
artemisinin

53
nicotine

54
penicillin

FIGURE 2.8 Examples of natural products with variable degrees of structural complexity.

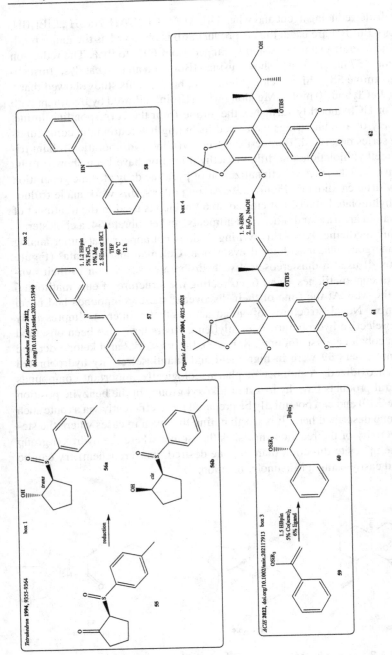

FIGURE 2.9 Trifluorothiomethyl derivatives of marketed pharmaceuticals.

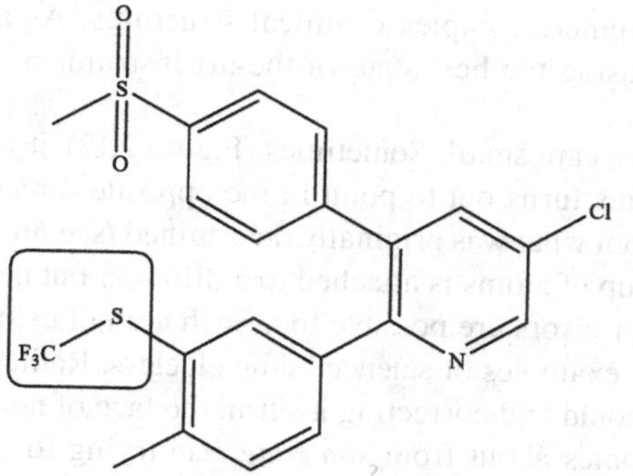

63
etoricoxib derivative

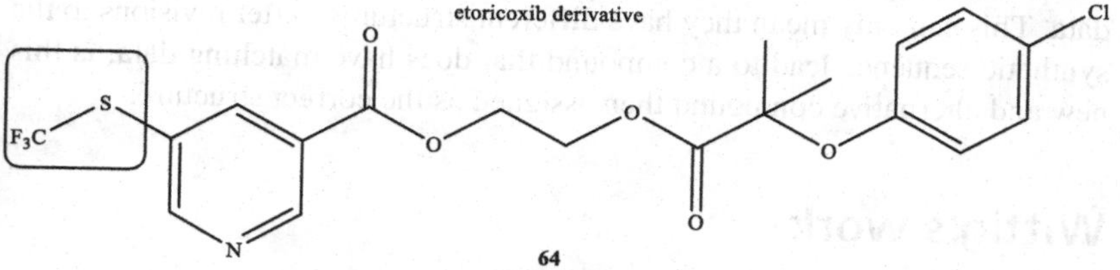

64
etofibrate derivative

FIGURE 2.10 Marketed pharmaceuticals with a hydroxy or alkoxyl moiety in the benzylic position.

A brief aside is appropriate here about correcting the reported structure of a compound, something that happens from time to time with the structure of natural products and before that, what a natural product is. A natural product (Figure 2.11) is a compound that is made by some sort of living organism including plants, fungi, molds, bacteria, and sponges, among others. Often, these compounds are nothing more than secondary metabolites of the source meaning that they have no discernible biological function for the organism making it; though sometimes they have functions related to insects. These compounds have a wide range of relevant functions in humans (and often other animals) ranging from cancer-curing to antiviral to antibiotic to hallucinogenic to lethal, and just about everything in between. These compounds are not made *for these reasons*; we simply noticed and took advantage. The isolation of these compounds and their structure elucidation is an active and exciting area of research. Since harvesting the source of these important compounds often requires activities like scuba diving or exploring other exotic locations, the lines of work and vacation may appear blurred to the onlooker. The antibiotic penicillin **54**, the antimalarial wonder drug artemisinin **52**, and even nicotine **53** are all examples of natural products. Many of these compounds have extremely complex chemical structures. As a result, even the best researchers using the best state-of-the-art instruments occasionally get something wrong.

Often, the mistakes are small. Sometimes (Figure 2.12), it will be that an atom or group of atoms turns out to point in the opposite direction in three-dimensional space from what was originally determined (see **66** vs. **67**). Other times, a group or group of atoms is attached to a different but nearby location in the molecule. Other errors are possible too. Such are not examples of misconduct, nor are they examples of science being clueless. Rather, it is science working the way it should and correcting itself in the face of new data. Often, this new evidence comes about from *someone else* trying to synthesize the compound, only to find that their instrumental data do not match the original data. This can only mean they have different structures. After revisions to the synthetic sequence lead to a compound that does have matching data, is this new and alternative compound then assigned as the correct structure.

Wittig's work

Brown's chemistry, though broad and synthetically useful, primarily involves the transformation of one functional group into another, though there are a few that generate carbon–carbon bonds. Wittig's chemistry converts aldehydes and ketones into olefins—alkenes—thus always making a carbon–carbon bond. And in fact, it is not just any carbon–carbon bond but a double bond, which in turn is able to be converted into a wide range of other functional groups though variants exist leading to carbon–heteroatom bonds. Moreover, conditions have been optimized to give control over the formation of a Z or E double bond.

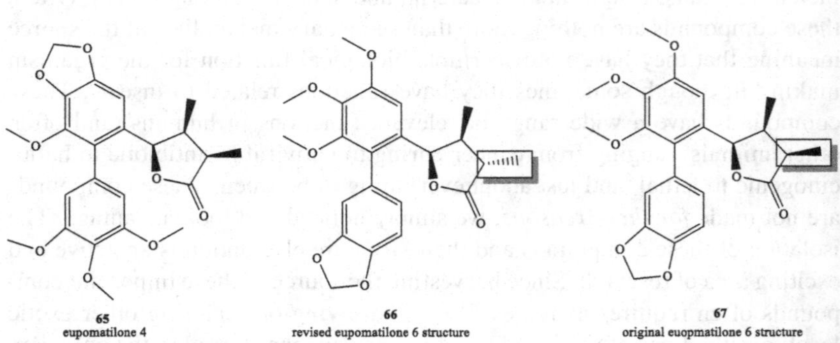

65
eupomatilone 4

66
revised eupomatilone 6 structure

67
original euopmatilone 6 structure

FIGURE 2.11   Uses of hydride reduction or hydroboration and subsequent chemistry in the synthesis of important products.

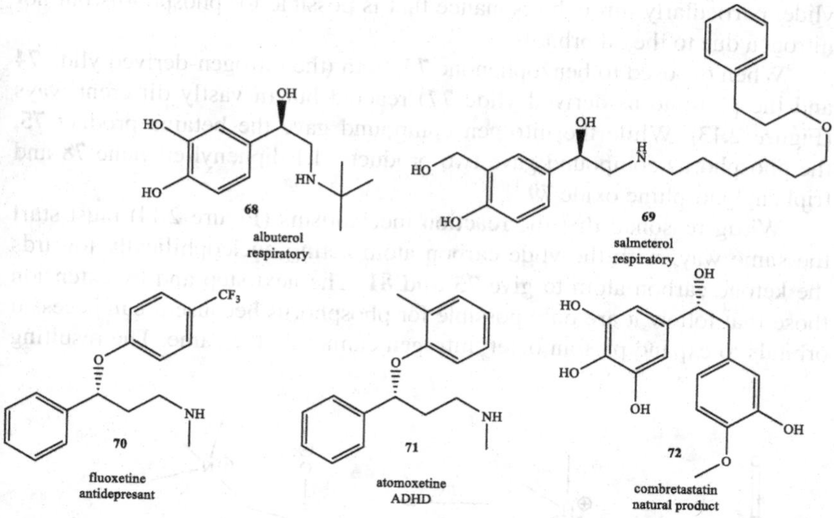

FIGURE 2.12 Structures of eupomatilone 4, the precursor to eupomatilone 6, and the original structure assignment of eupomatilone 6.

George Wittig (and remember, it is pronounced Vittig; he's German), the other half of the 1979 Nobel Prize in chemistry, was awarded a share for his work using phosphorus compounds. In particular, the Wittig reaction is one that allows for the facile formation of alkenes from ketones or aldehydes.

Wittig and his coworkers somewhat stumbled upon the ylide, the key intermediate in the Wittig reaction. Their initial goal was an "absurd experiment" (his words, not mine) involving an attempt to generate a pentavalent nitrogen. When this (of course) failed, they moved to arsenic, antimony, bismuth, and phosphorus, a natural move given they all are in the same family on the periodic table. Prior, it was found by Wittig that methyl and phenyl lithium will deprotonate a tetra-alkyl ammonium salt, resulting in a species where an anion is on the carbon atom α to (attached to) the nitrogen atom, which as an ammonium nitrogen was positively charged. They coined the term ylide because the bonding of the carbon mentioned is homopolar (yl) and ionic negative specifically—ide) at the same time.

The organophosphorus compounds central to what would become known as the Wittig reaction were also found to undergo deprotonation upon treatment with organolithium reagents but more readily than their nitrogen counterparts. Since nitrogen and phosphorus, again, are in the same group/family on the periodic table, their similar chemical behavior is expected and with phosphorus's access to d orbitals, it is also logical that these systems would be more readily deprotonated on account of enhanced stability of the ylide, particularly through resonance that is possible for phosphorus but not nitrogen due to the 3d orbitals.

When exposed to benzophenone **73**, both (the nitrogen-derived ylide **74** and the phosphorus-derived ylide **77**) reacted but in vastly different ways (Figure 2.13). While the nitrogen compound gave the betaine product **75**, the phosphorus compound gave two products: 1,1-diphenyl ethylene **78** and triphenylphosphine oxide **79**.

Wittig reasoned that the reaction mechanisms (Figure 2.14) must start the same way; with the ylide carbon atom acting nucleophilically towards the ketone carbon atom to give **75 and 81**. The next step and by extension those that follow it are only possible for phosphorus because it can access d orbitals to expand past an octet; nitrogen cannot do the same. The resulting

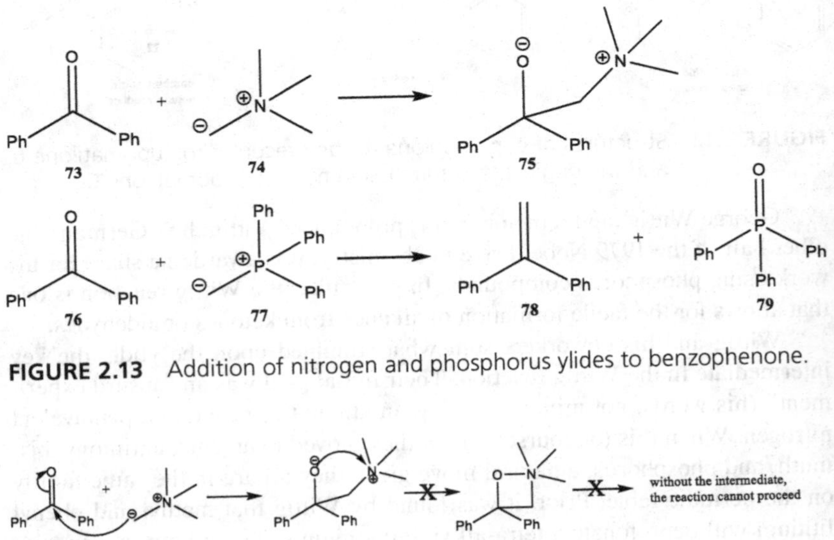

FIGURE 2.13 Addition of nitrogen and phosphorus ylides to benzophenone.

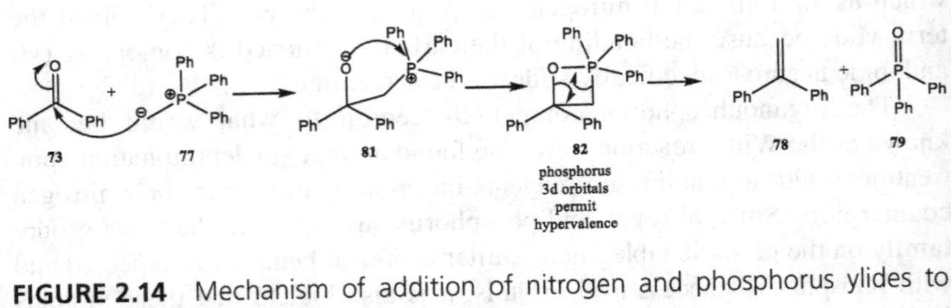

FIGURE 2.14 Mechanism of addition of nitrogen and phosphorus ylides to benzophenone.

four-member ring **81** then collapses to give both the alkene **78** and phosphine oxide **79** products.

More complex systems can form either the more or the less sterically hindered alkene product. The Wittig reaction gives a range of selectivity with respect to the stereochemical configuration of the alkene (i.e., E vs. Z). Each can be selected for and a mixture of the two may also be produced. Although conclusive explanations for the stereochemical selectivity are still somewhat lacking, there are several ideas that are beyond the scope of this discussion. The following trends, at least, are clear:

1. Ylides containing stabilizing groups or formed from trialkyl phosphines tend to give the E product
2. Ylides lacking stabilizing groups and formed from triaryl phosphines tend to give the Z product or a mixture

The Schlosser modification[9] of the Wittig olefination uses an excess of lithium salts and a lithium base such as an organolithium. The role of lithium appears to be to stabilize the betaine intermediate—the intermediate formed before the cyclization to the oxaphosphetane. Using a sterically hindered proton source (such as *tert*-butanol), the *trans* betaine can be formed preferentially prior to the formation of the oxaphosphatane. This trans relationship is then preserved in the oxaphosphatane intermediate and later the olefin as the E product.

Not all organophosphanes, however, are appropriate Wittig reagents (Figure 2.15). Although in truth, any tetravalent phosphorus with at least one α-hydrogen-bearing alkyl group attached can function as a Wittig reagent, in practice, there are such things as **bad** Wittig reagents; reagents where there are different alkyl groups that each have an α-hydrogen atom.

Although in the case of the "bad" reagent **83**, anion stability and steric hindrance may allow for some selective deprotonation, a mixture will almost

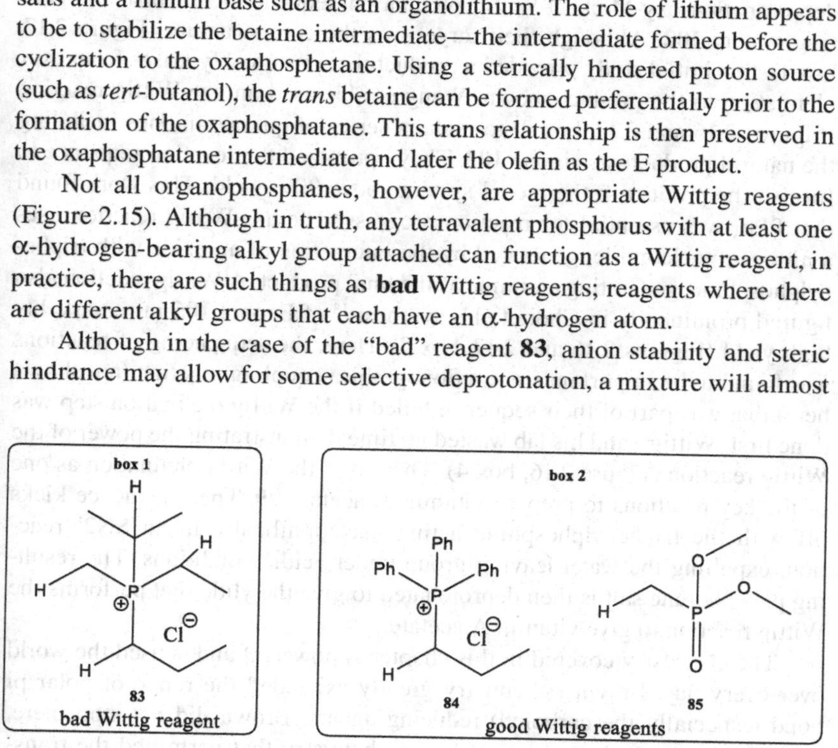

FIGURE 2.15 Examples of good and bad Wittig reagents.

inevitably be observed. In each of the "good" cases **84 and 85**, only one position can be deprotonated and thereby converted into a nucleophile. Thus, reagents structured like these are a far better choice for synthesis.

MODIFICATIONS

A modified version of the reaction employs phosphonate esters, rather than phosphoranes. This version has several names but will be referred to as the Horner–Wadsworth–Emmons reaction here. The ylides formed from these reagents tend to be more reactive and, thus, will even cause the olefination of less reactive ketones that resist the phosphorane derivatives.[10]

In their pursuit of the synthesis of fargenone and the fargenin family of natural products Figure 2.16, box 1), Denton and Scragg[11] employed a deprotonation-oxy-Michael-Wittig olefination sequence to furnish the core ring system **102**. Although their target was the core structure (Figure 2.17, box 1), this initial study provided a product that presumably arises from over olefination to the product shown. Wang et al.[12] used an aza-Wittig reaction (Figures 2.16 and 2.17, box 2) to make a series of quinazolinones including the natural product vasicinone **103**. They optimized the reaction sequence to furnish the product in the model system above 90% yield. This work found that PPh$_3$ was essential for reaction progress. The aza-Wittig reaction uses a nitrogen nucleophile, generated by the interaction of an azide with triphenylphosphine, generating an imine as the final product. A Wittig reaction also figured prominently in Choi et al.'s synthesis[13] of Egonol **105** in remarkably high yield (Figures 2.16 and 2.17, box 3). Here, the sequencing of reactions was found to be important since a Sonogashira coupling reaction (not shown here) that was part of their sequence failed if the Wittig olefination step was done first. Wittig[14] and his lab wasted no time demonstrating the power of the Wittig reaction (Figure 2.16, box 4). They used the Wittig olefination as one of the key reactions to prepare vitamin A acetate **99**. Their sequence kicks off with the triphenylphosphine acting nucleophilically in an SN2' reaction, expelling the water-leaving group under acidic conditions. The resulting phosphorane salt is then deprotonated to give the ylide that performs the Wittig reaction to give vitamin A acetate.

The chemistry covered in this chapter is powerful and is used the world over every day. Brown's chemistry greatly expanded the range of polar pi bond (especially the carbonyl) reducing agents. Brown did not stop there, exploring a tremendous range of boron chemistry that permitted the transformation of the alkene into a wide range of other functional groups. Wittig,

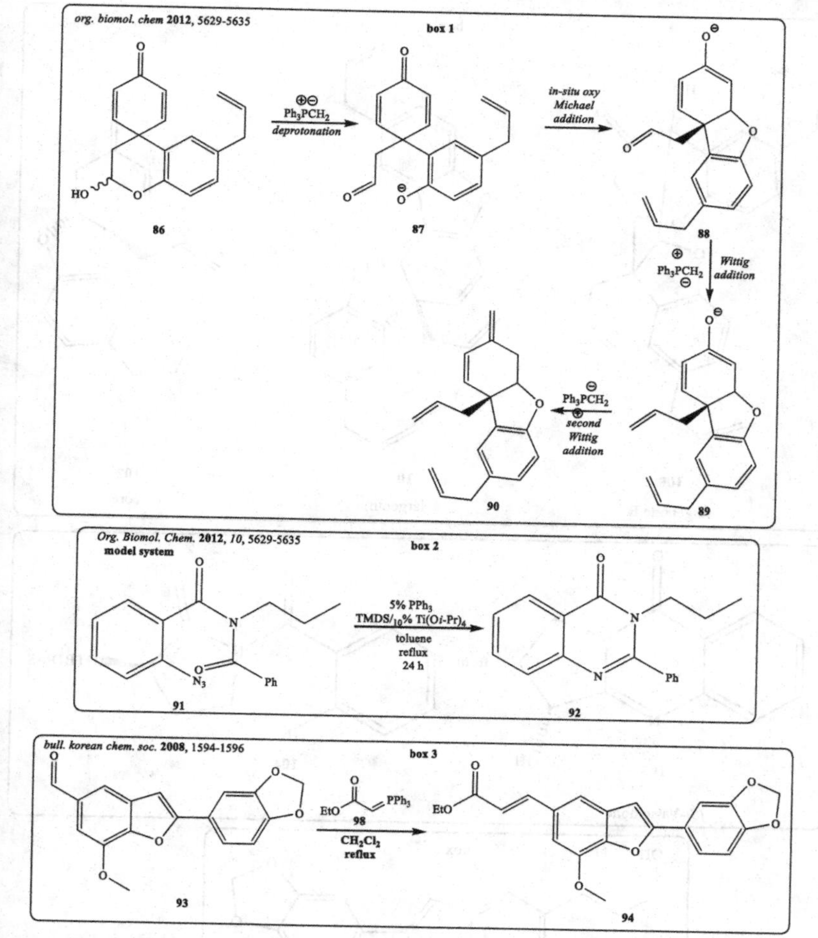

FIGURE 2.16 Examples of the use of the Wittig reaction in the synthesis of natural products.

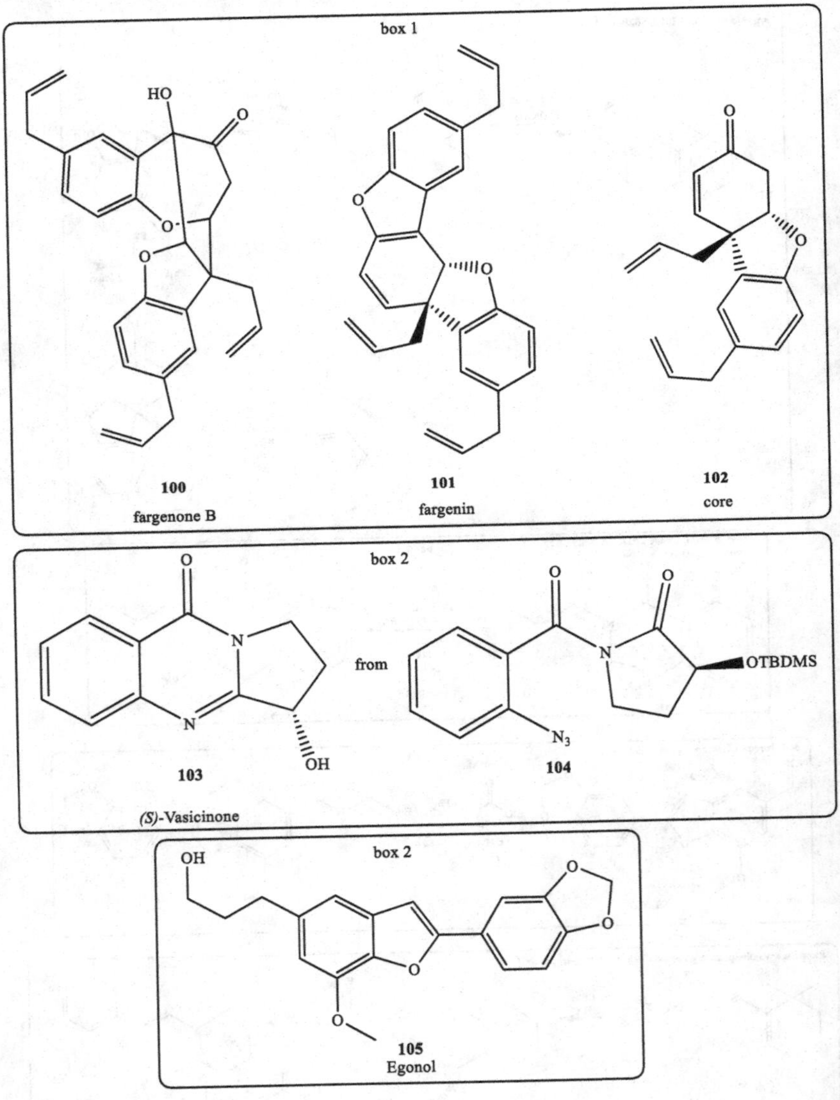

FIGURE 2.17 Examples of natural product targets prepared using the Wittig reaction in at least one of the steps in the chemical sequence.

meanwhile, developed chemistry that converts the carbonyl into an alkene. These three families of transformations greatly expanded the synthetic toolbox of synthetic organic chemists. It is almost impossible to overstate the combined contribution to synthetic chemistry.

GENERAL REFERENCES

https://www.nobelprize.org/uploads/2018/06/brown-lecture.pdf, last checked 11/7/22.

Smith, Michael B. *Organic Synthesis*, second edition, McGraw Hill, 2002.

Smith, Michael B. *March's Advanced Organic Chemistry*, seventh edition, Wiley, 2013.

NOTES

1 officialdata.org, last checked 11/7/22.

2 Bueno, A. B.; Carreño, M. C.; García Ruano, J. L; Peña, B; Rubio, A.; Hoyos, M. A. *Tetrahedron* **1994**, *50*, 9355–9364.

3 Lei, S.; Pan, T.; Wang, M.; Zhang, Y. *Tetrahedron Letters* **2022**, 153949 (https://doi.org/10.1016/J.tetlet.2022.153949), last checked 11/7/22.

4 Zhou, X. Y.; Zhang, M.; Liu, Z.; He, J.-H.; Wang, X.-C. *Journal of the American Chemical Society* **2022**, *144*, 14463–14470.

5 Coleman, R. S.; Gurrala, S. R. *Organic Letters* **2004**, *6*, 4025–4028.

6 Carroll, A. R.; Taylor, W. C. *Aust. J. Chem* **1991**, *44*, 1615.

7 Dong, W.; Ye, Z.; Zhao, W. *Angewandte Chemie International Edition* **2022**, *61*, e202117413 (https://doi.org/10.1002/anie.202117413), last checked 11/7/22.

8 Mitsunobu, O.; Yamada, Y. *Bulletin of the Chemical Society of Japan.* **1967**, *40*, 2380–2382.

9 Schlosser, M.; Christmann, K. F. *Ang. Chem. Int. Ed. in English*, **1966**, *5*, 126.

10 Smith, Michael B. *March's Advanced Organic Chemistry*, seventh edition, Wiley, 2013.

11 Denton, R. M.; Scragg, J. T. *Organic and Biomolecular Chemistry* **2012**, *10*, 5629–5635.

12 Wang, L.; Wang, Y.; Chen, M.; Ding, M.-W. *Advanced Synthesis and Catalysis* **2014**, *356*, 1098–1104.

13 Choi, D. H.; Hwang, J. W.; Lee, H. S.; Yang, D. M.; Jun, J.-G. *Bull. Korean Chem. Soc.* **2008**, *29*, 1594–1596.

14 https://www.nobelprize.org/uploads/2018/06/wittig-lecture.pdf, last checked 11/7/22.

15 Acid chlorides react with hydroxylic solvents such as alcohols. To get a reduction as the major pathway, an aprotic solvent must be used.

1981—
Fukui and
Hoffmann

3

The 1981 Nobel Prize in Chemistry was awarded to Kenichi Fukui and Roald Hoffmann for their work, developed independently, on the course of chemical reactions. At the most basic level, these theories provide a molecular orbital-level rationalization for the course of pericyclic and electrocyclic chemical reactions in the Woodward and Hoffmann case and all reactions except for nuclear chemistry in Fukui's. Woodward and Hoffmann define electrocyclic transformations as the formation of a single bond between the termini of a linear system containing k–π electrons and the converse process. These electrocyclic reactions are part of a larger class of reactions called pericyclic reactions. The Oxford chemistry dictionary[1] defines these as a type of concerted chemical reaction that proceeds through a cyclic conjugated transition state. This larger class of pericyclic reactions also includes cycloadditions such as the Diels–Alder reaction and cheletropic reactions. They also include both the Cope and sigmatropic rearrangements and the ene reaction.

As there is no singular chemical reaction involved with this Nobel award, the understanding of the course of chemical reactions (the material in this award) enhanced control over several chemical transformations. It is worth stressing for the less experienced reader that a better understanding of the mechanisms of chemical reactions permits a better understanding of how to modify the chemical conditions to improve or even specify product formations. The range of reactions these rules govern is wide. The portion of the award given to Hoffmann covers reactions referred to as sigmatropic reactions—of which there are many—while the portion awarded to Fukui applies to virtually every chemical reaction except for nuclear chemistry. Thus, an exhaustive discussion of these reactions will not be done here; they could fill an entire volume on their own.

Before going into their science, one of the scientists, Roald Hoffmann, must be discussed at least briefly.[2] Hoffmann was born in what was then

DOI: 10.1201/9781003006848-3

the Second Polish Republic, now part of what is currently Ukraine.[3] When he was five, the Nazis showed up and Roald along with his family—being Jewish—were rounded up into a labor camp. After bribing a guard, Roald, his mother, two uncles, and an aunt were whisked away to the attic and storeroom of a local schoolhouse. This was "home" for 18 months, until Roald was seven. His father, who remained behind in the camp, was eventually tortured and killed for his involvement in a plot to arm the other prisoners at the camp. Save for one grandmother and a few others, along with those who initially escaped with Roald, most of his family perished during the Holocaust. His story is exceptionally inspirational and is one of five Nobel Prize winners I have seen give a lecture and the only one I have shaken hands with, something that was—more for his personal story than his scientific—a very humbling experience. He is also, in my opinion, an outstanding speaker.

So, if not a chemical reaction, what is this award for? To fully understand, at least some mention of orbitals must be discussed, specifically molecular orbitals and this is where Fukui's work starts. Chemical reactions—excluding nuclear reactions—are governed by what are called HOMO–LUMO interactions. These are the interactions between the Highest Occupied Molecular Orbital (where the electrons **are**) and the Lowest Unoccupied Molecular Orbital (where electrons **are *not***). The HOMO of one molecule or atom and the LUMO of another molecule or atom (or two different regions of the same molecule during some intramolecular reactions) are ordinarily closest to one another in energy. These are referred to as frontier molecular orbitals.

Given that the HOMO is (essentially) where the electrons reside and the LUMO is an absence of electrons, it makes sense that the two would interact with each other to form bonds such as a covalent bond (a stable one, anyway); the sharing of a pair of electrons. This is especially true if one considers the alternative interactions:

> LUMO-LUMO interactions-unlikely to produce a bond since completely lacking electrons, there is no way for such an interaction to form a bond, which recall is the sharing of electrons.
>
> HOMO-HOMO interactions-unlikely since this effectively results in an overload of electrons in any δ-bond system as bonds are the sharing of two electrons. An interaction between two HOMOs places electrons in not just a bonding orbital but also an antibonding orbital.

As will be seen later in this chapter, these HOMO–LUMO interactions will control pericyclic reactions, which is Hoffmann's part of this year's prize. As a set of straightforward examples, here, the orbital interactions during an S_N2 and E2 reactions and the trapping of a carbocation by a nucleophile will be considered, for a novice reader's familiarity's sake.

THE S$_N$2 AND E2 REACTIONS

Recall from any standard introductory organic chemistry course that the S$_N$2 reaction gives the stereochemistry-inverted product *via* a process commonly referred to as backside attack, so called because the nucleophile approaches the substrate **1** from the opposite direction that the leaving group (LG) is pointing in in three-dimensional space as shown in Figure 3.1. This occurs due to the HOMO–LUMO interactions shown in Figure 3.2. Here, the nucleophile (Nu:)—which donates electrons—is providing the HOMO because to donate electrons, you must have an orbital that is occupied by electrons; you cannot donate electrons from an empty orbital (the LUMO). Meanwhile, the LUMO in this case is the antibonding orbital for the carbon-LG bond. Placing electrons into this orbital destabilizes the carbon-LG bond and in fact any bond would behave the same way. This ultimately explains several aspects of this reaction. First, it explains why it is a one-step process; that is, why the nucleophile–carbon bond is formed concurrently with the carbon-LG's bond destruction or rupture. Second, it explains the direction of the attack; it explains why the nucleophile must approach from the opposite direction since this is the direction the antibonding orbital is facing. The antibonding orbital, to keep the electrons in the bonding orbital from overlapping with it and causing the bond to destroy itself, is pointing in the opposite direction from the bond. This means that the LUMO which the HOMO must interact with is pointing in the direction of the "backside" of the molecule, relative to that of where the LG is pointing. Therefore, the nucleophile **must** approach the molecule from this direction for this reaction to occur. As the electrons in the HOMO of the nucleophile are deposited into the carbon-LG bond's antibonding orbital (the LUMO), the carbon-LG bond must be broken; however, "made" the carbon–nucleophile bond is during the reaction is the same degree to which the carbon-LG bond has been broken.

Also, recall that the E2 reaction behaves similarly. In this case, it is not a nucleophile that deposits electrons into the antibonding orbital of the carbon-LG bond but the electrons in the carbon-hydrogen bond on the neighboring

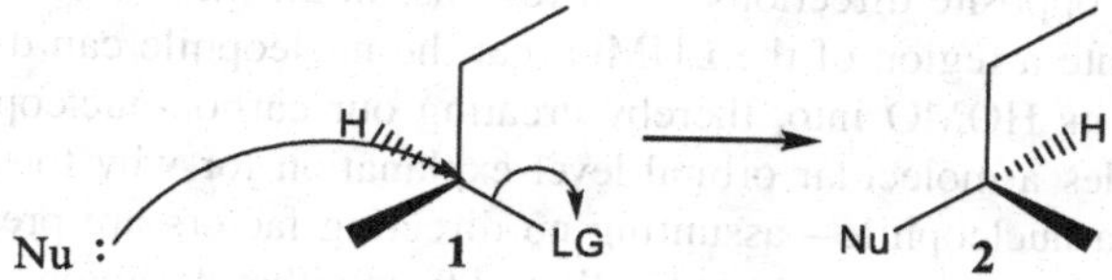

FIGURE 3.1 Depiction of backside attack in S$_N$2 reaction.

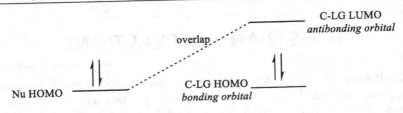

FIGURE 3.2 HOMO–LUMO interactions in S_N2 reaction.

carbon atom. This is precisely why it is specifically a hydrogen atom anti—pointing in the opposite direction—to the LG that is removed during this reaction. This hydrogen and only this hydrogen's C-H bond and the electrons comprising that bond are in the necessary position to overlap with the carbon-LG bond's antibonding orbital. Now, in the case of this E2 reaction, the double bond is however made as the carbon-LG bond is broken.

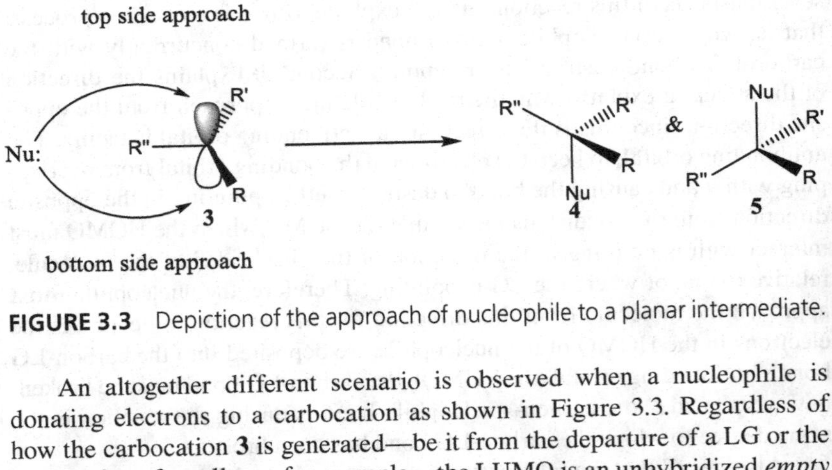

FIGURE 3.3 Depiction of the approach of nucleophile to a planar intermediate.

An altogether different scenario is observed when a nucleophile is donating electrons to a carbocation as shown in Figure 3.3. Regardless of how the carbocation **3** is generated—be it from the departure of a LG or the protonation of an alkene, for example—the LUMO is an unhybridized *empty* p orbital. On an XYZ axis, this unhybridized and empty p orbital extends above and below the plane of the molecule as drawn. Other orientations would place it in front of and behind or to the left and the right, essentially any pair of opposite directions. Both regions, in all these cases, are empty and constitute a region of the LUMO that the nucleophile can donate electrons from its HOMO into, thereby creating our carbon-nucleophile bond. This provides a molecular orbital-level explanation for why the capture of cations by a nucleophile—assuming no directing factors are present in the molecule—can happen in equal likelihood from either direction, resulting in a mixture of stereoisomers **4 and 5** as the product of a reaction.

This finishes the part of the prize awarded to Fukui; the contribution that identifies the orbitals responsible for the chemical reactions observed. The work of Woodward and Hoffmann for which the latter was awarded the Nobel Prize analyzes specific HOMO–LUMO interactions and uses them to describe the stereochemical outcome of a class of reactions known as electrocyclic reactions. Most of the reactions covered in this section on Hoffmann's work are remarkable for several reasons; two, however, stand out as noteworthy here: 1) tremendous stereo control and 2) wide functional group tolerance. To the inexperienced reader, these combined, at the most basic level means that one "handed (right or left)" molecule can be made or at most 2, in preference over several other options and point 2 means that other portions of the molecule do not also react under the conditions. The importance of the first point should be easy to see value in if looked at from the point of view of lowering the number of products generated, which such selectivity literally is. The second point's importance will be further discussed in the next chapter of this volume.

Hoffmann's portion of the award is awarded for the development of what is now known as the Woodward–Hoffmann rules, first articulated in 1965[4] which explain the course of electrocyclic reactions like those shown in Figure 3.4. These processes can occur either by disrotatory **6a and 6b →** **7a and 7b** or conrotatory motions **8a and 8b → 9a and 9b** of the reacting carbon atoms during electrocyclic reactions. Disrotatory motion is when each atom rotates in opposite directions (on a clock) while conrotatory motion rotates them in the same direction. This might seem counterintuitive since

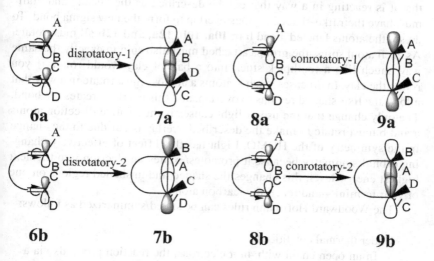

FIGURE 3.4 Orbital symmetry diagram for disrotatory and conrotatory ring-closing reactions.

the disrotatory motion looks like they are rotating towards each other and the conrotatory motion looks as if they are rotating away from each other. Rather than this sort of rotation giving rise to the label, it is all about how each rotates with respect to a clock.

Although these reactions occur thermally as well as photochemically, they always proceed diastereospecifically, rather than diastereoselectively. At worst, a mixture of two stereoisomers, each arising from either a conrotatory or disrotatory spin **7a versus 7b and 8a versus 8b**—just in opposite directions—is made. Often, this results in a pair of enantiomers though if there are other pre-existing stereocenters in the starting material, diastereomers can also be formed.

Woodward and Hoffmann offered as the explanation for these cumulative observations that these stereochemical results were due to the orbital symmetry present in the HOMO of the open-chain partner in all these reactions. Under photochemical conditions, an electron is excited into a higher energy level, causing the identity of the HOMO to change. This likewise changes the symmetry of the HOMO, and this change is what leads to the different reaction courses of conrotatory versus disrotatory. Figure 3.5 compares the electrocyclic formation of cyclobutene and cyclohexene from their respective conjugated open-chain partners under both thermal and photochemical conditions. It should be noted that specific wavelengths of light are necessary for excitation, not just *any* light will do in most cases. Notice that in these electrocyclic reactions, it is the HOMO that reacts and that it is reacting in a way that can be described as the "head" and "tail" must have their like-shaded regions overlap to form the new sigma bond. To do so, the atoms labeled a and b in **10a, 10b, 12a, and 12b** all must rotate. As each atom spins, the groups attached must also spin in exactly the same way, much like a toothpick stuck into a ball of clay would rotate if you rotate the clay. In all cases, each—atoms a and b—must rotate in a way that allows the like-shaded regions to overlap, and this overlap creates the bond. The only change that the use of light causes is the relative direction atoms a and b must rotate to make the described overlap occur due to the change in the symmetry of the HOMO. Light has this effect of effectively changing orbital symmetry because it promotes/excites an electron into the next highest energy level; this changes the shaded and unshaded regions on one of your termini—and reacting—carbon atoms.

The Woodward Hoffmann rules can be stated/summarized as follows:

Under thermal conditions
 In an open chain with 4n π electrons, the reaction proceeds via a conrotatory process
 In an open chain with 4n + 2 π electrons, the reaction proceeds via a disrotatory process

Under photochemical conditions
 In an open chain with 4n + 2 π electrons, the reaction proceeds via
  a conrotatory process
 In an open chain with 4n π electrons, the reaction proceeds via a
  disrotatory process

Each of these conditions is required in their respective cases because of the orbital overlap requirements to form a bond. Only via these specific rotating motions can the necessary overlap occur as demonstrated in Figure 3.5. As the reacting atoms rotate to cause the required overlap, all attached substituents must rotate with it, directing the attached groups into the relative positions

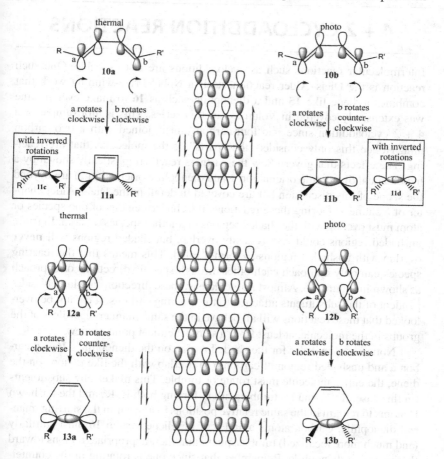

FIGURE 3.5 Comparison of orbital symmetry under thermal and photo conditions.

observed. Recall, as previously discussed, the inverse of each of the rotations can occur in all cases, as long as the conrotatory or disrotatory nature is preserved. A subtlety should not be overlooked here. This strict adherence to the rotation rules is the genesis of the stereopecificity. That is, notice how if **10a** were to rotate such that atoms a and b were to rotate counterclockwise instead of both clockwise, a different trans product **11c** would still be formed. On the other hand, if **10b** similarly inverted its rotation with atom a rotating counterclockwise and atom b rotating clockwise, **11d**, a cis product and the enantiomer of **11b** would be formed. In short, changing the actual direction of the rotation *does not* change the relative orientation of the substituents in the product.

4 + 2 CYCLOADDITION REACTIONS

Intermolecular reactions such as cycloadditions are also possible. One such reaction is the Diels–Alder reaction, itself a Nobel Prize-winning work that combines a diene like **15** and a dienophile such as **16** to make cyclohexane was extensively covered in Volume 1 of this series. This is an example of a 4 + 2 cycloaddition since one four-carbon unit is joined with a two-carbon unit. Notice this only considers the "parts" of the molecules that are reacting. The effects that govern how the partners react (simplistically shown by a hypothetical dienophile orientation 1 and dienophile orientation 2) are beyond the scope of this discussion but are covered in detail in the Diels–Alder chapter of Volume 1. During these reactions, the shaded regions of one species or atom must overlap with the shaded regions of another species or atom. In truth, unshaded regions could just as easily overlap, but shaded regions will **never** overlap with unshaded regions to form a bond. This means that the reacting species can only approach each other through specific directions of approach as shown in Figure 3.6. Almost miraculously, these directions are largely independent of the substituents attached to the reacting carbons. It cannot be overlooked that these reactions will all behave in the same manner regardless of the groups in the molecule, at least from a stereochemical point of view.

Notice how in order for the shaded region on the dienophile's **16a–d** carbon a and unshaded region of carbon b to overlap with the like regions on the diene, the entire molecule must rotate or tumble. This makes any substituents (in this case, R, R′, and H) tumble with it, forcing both R, R′, and the unshown H atoms to remain in the same relative positions that occur in the starting material dienophile. Carbon atoms a and b of the diene system **16** must similarly (and much more localized) tumble/spin to aim their appropriate regions toward the incoming dienophile. Remember that since one is rotating in the counterclockwise direction and the other in the clockwise direction, this is a disrotatory process. Notice how this causes both attached methyl groups, which are

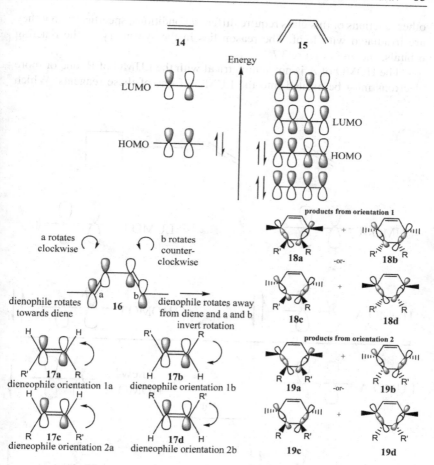

FIGURE 3.6   depiction of how different stereoisomers can be formed based on orbital symmetry.

pointing "out" of the diene molecule shown here rotate in such a way that they end up on the same side as each other in the product.

2 + 2 CYCLOADDITION REACTIONS

There are also 2 + 2 cycloadditions. Whereas the 4 + 2 version has four atoms in one π system and two atoms in the other, these 2 + 2 cycloadditions have two atoms in both. These 2 + 2 cycloadditions, however, unlike most of the

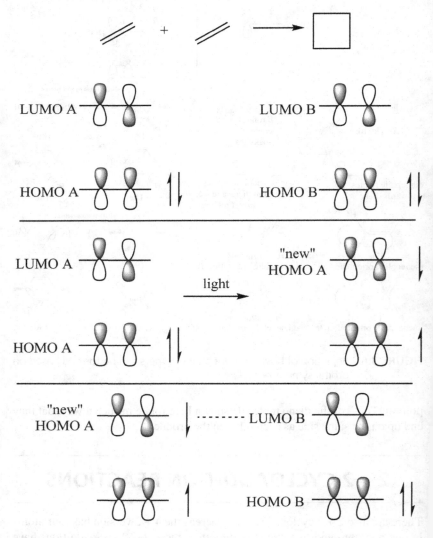

other reactions of this class require different conditions, specifically that they are irradiated with light. The reason lies in the symmetry of the reacting orbitals, shown in Figure 3.7.

The HOMO of A is not symmetrical with the LUMO of B, one or more electrons must be excited into the LUMO of one of these reagents. Which

FIGURE 3.7   HOMO–LUMO interactions for 2 + 2 cycloadditions.

reagent sees this "excitement" is dependent on the functional groups attached and, it is only one of the reagents that is so activated. A look at the corresponding HOMO–LUMO diagram shows why this solves the problem, assuming "A" is activated.

Because an electron has been excited into the LUMO of A, this molecular orbital is no longer a LUMO as it is no longer unoccupied. The orbitals (LUMO B and HOMO A* ("new" HOMO A) as shown in this general example) are now able to overlap and form the cycloaddition product.

SIGMATROPIC REARRANGEMENTS

Like the electrocyclic and cycloaddition reactions, sigmatropic rearrangements are concerted reactions. Unlike the other systems, this reaction will not only make but *break* a sigma bond as shown in Figure 3.8. The following changes happen during these types of reactions:

> A group attached by a sigma bond migrates to the end of an adjacent pi system. This breaks the sigma bond while making another sigma bond. A simultaneous shift of pi electrons occurs.

Similar to how cycloadditions are given categories such as 2 + 2 and 4 + 2, sigmatropic rearrangements have their own classification system. The variables used to describe these systems are the number of groups in the migrating fragment (i) and the pi system (j) that are directly involved in the bonding changes and are given in a [i, j] format. Additional new terms are used to describe the stereochemical outcome of these transformations—suprafacial and antarafacial. In a suprafacial process, the migrating group remains associated with the same face of the pi system while in antarafacial processes, the migrating group moves to the opposite face of the pi system. Examples of these reactions are shown in Figure 3.9.

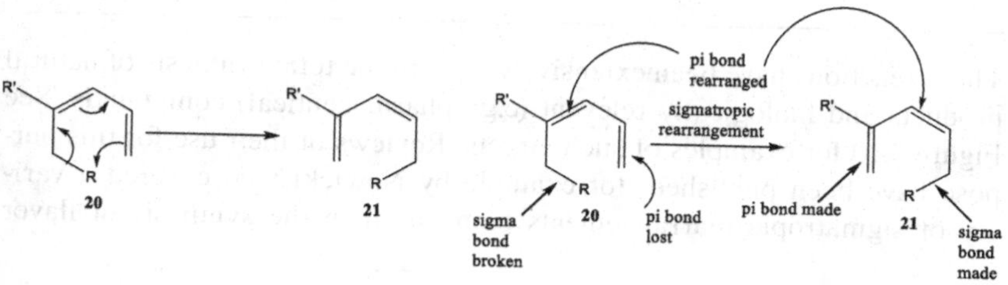

FIGURE 3.8   Arrow-pushing reaction mechanism for sigmatropic rearrangement and explanation of bond formation/rupture.

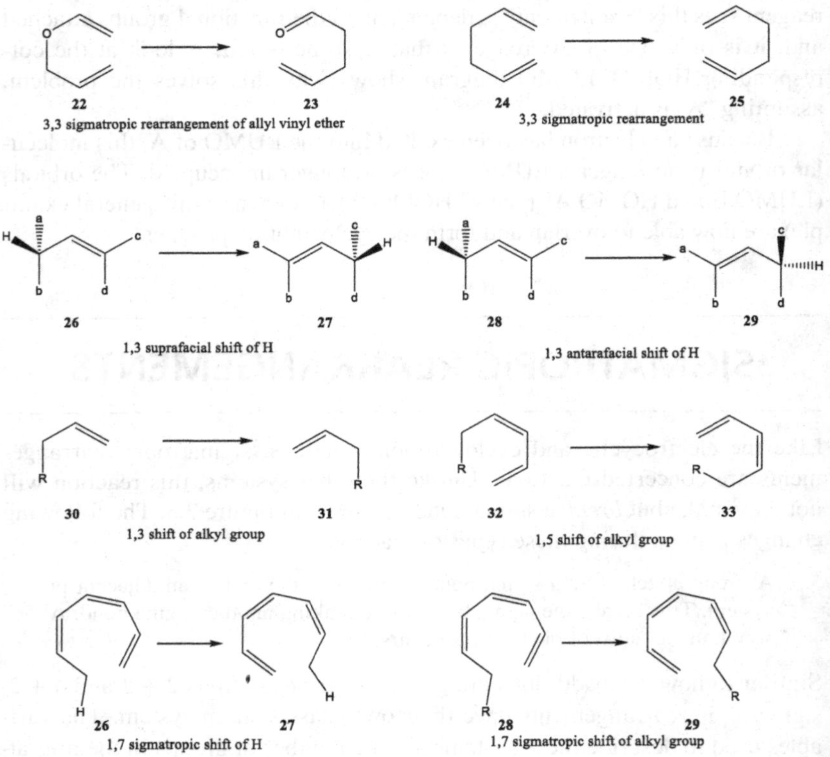

FIGURE 3.9 Examples of sigmatropic rearrangements.

APPLICATION TO THE SYNTHESIS OF IMPORTANT MOLECULES

These reactions have been extensively used in the total synthesis of natural products and biologically relevant (e.g., pharmaceutical) compounds. See Figure 3.10 for examples of such targets. Reviews of their use for this purpose have been published, for example by Nowicki who covered a variety of sigmatropic rearrangements being used in the synthesis of flavor

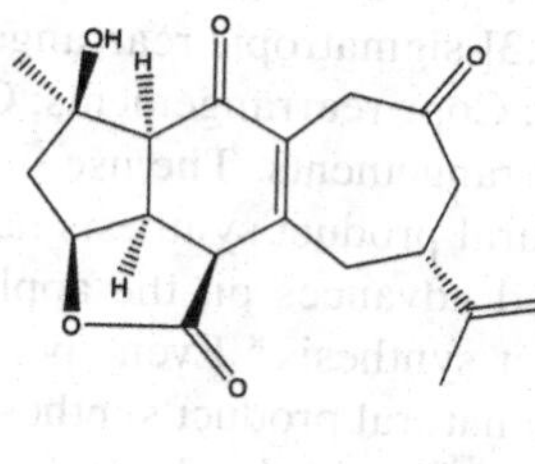

38
scabrolide A

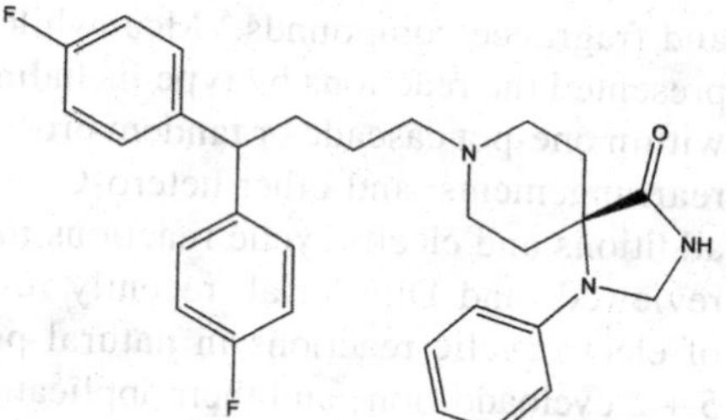

39
fluspirilene

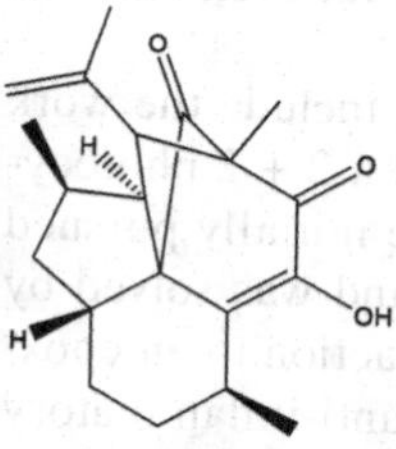

40
(-)-elisapterosin B

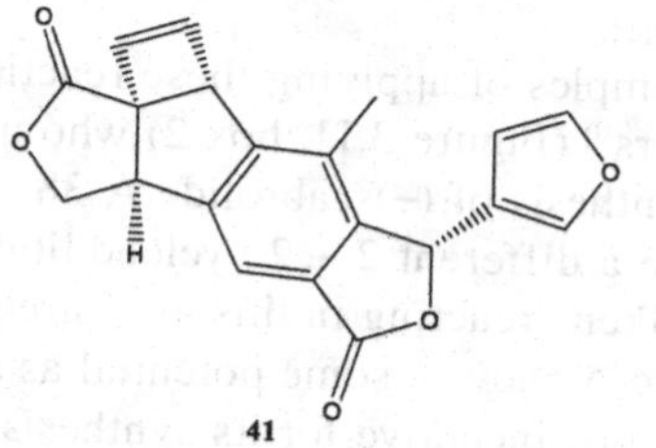

41
salvileucalin C

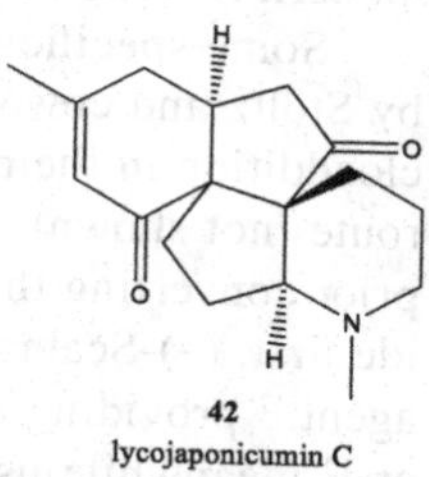

42
lycojaponicumin C

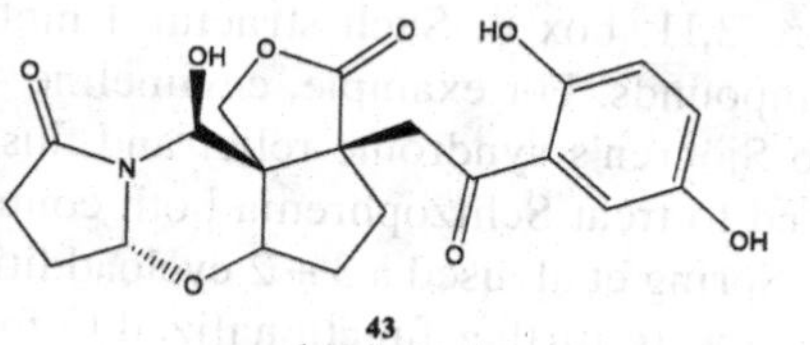

43
sinensilactam A

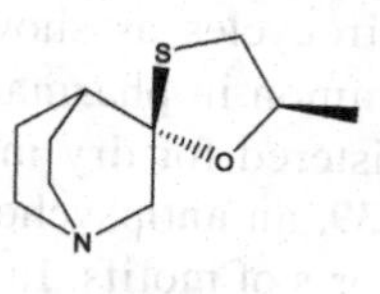

44
cevimeline

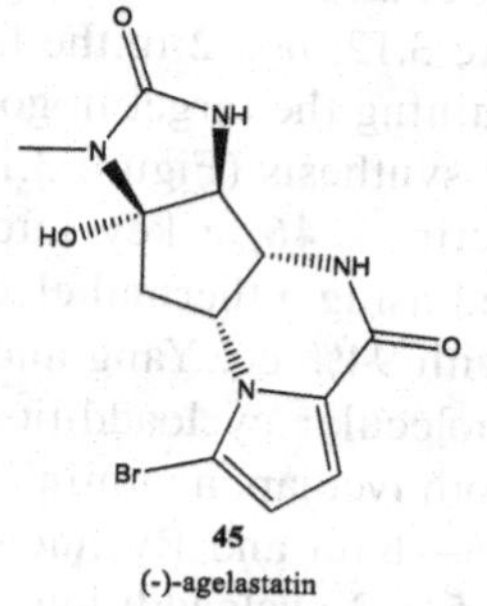

45
(-)-agelastatin

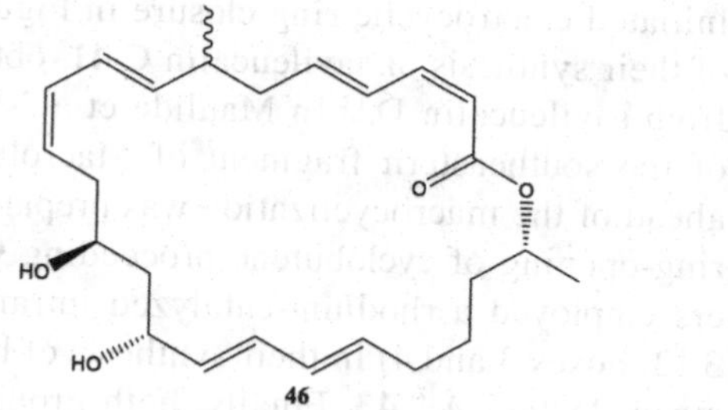

46
Macrolactin A

FIGURE 3.10 Natural products or other compounds of therapeutic interest where an electrocyclic or sigmatropic reaction figured prominently in the synthesis.

and fragrance compounds.[5] Meanwhile, Ilardi et al.[6] penned a review that presented the reactions by type including [3,3]-sigmatropic rearrangements within one-pot cascade or tandem processes; Cope rearrangements; Claisen rearrangements; and other hetero-Cope rearrangements. The use of cyclo-additions and electrocyclic reactions in natural product synthesis has been reviewed[7] and Ding et al. recently reviewed advances on the application of electrocyclic reactions in natural product synthesis.[8] Even specifically 5 + 2 cycloadditions and their application to natural product synthesis have been reviewed.[9] The use of this family of reactions is clearly powerful, far reaching, and common; it is nearly impossible to account for even only the modern reviews in detail.

Some specific examples of applying these reactions include the work by Stoltz and coworkers[10] (Figure 3.11, box 2) who used a 2 + 2 photocy-cloaddition in their synthesis of (−)scabrolide A **38**. The initially pursued route (not shown) gave a different 2 + 2 cycloaddition and was solved by prior converting the alkene reacting in this undesired reaction to an epox-ide first. (−)-Scabrolide A showed some potential as an anti-inflammatory agent,[11] providing obvious incentive for its synthesis. Spring and coworkers[12] meanwhile used cycloaddition strategies to create diverse heterocy-clic spirocycles as shown in Figure 3.11, box 4. Such structural motifs are common in pharmaceutical compounds. For example, cevimeline **44**, administered for dry mouth due to Sjögren's syndrome relief and fluspi-rilene **39**, an antipsychotic drug used to treat Schizophrenia both contain these sorts of motifs. In their work, Spring et al. used a 3 + 2 cycloaddition to create a series of isoxazoles which were further functionalized to form spirocycles. Further examples include Ding et al.'s work in using a photo-initiated electrocyclic ring closure in Figure 3.12, box 2 in the final steps of their synthesis of savileucalin C **41**, obtaining the target in good yields from savileucalin D.[13] In Maulide et al.'s[14] synthesis (Figure 3.11, box 3) of the southeastern fragment of Macrolactin A **46**, a key intermediate ahead of the macrocyclization was prepared using a thermal electrocyclic ring-opening of cyclobutene proceeding with 94% ee. Yang and cowork-ers employed a rhodium-catalyzed intramolecular cycloaddition (Figure 3.12, boxes 3 and 4) in their synthesis of both lycojaponicumin[15] C **42** and sinensilactam A[16] **43**. Finally, both groups—Kim and Rychnovsky[17] and Harrowven[18] et al.—independently used a 5 + 2 cycloaddition as the last step in the synthesis of (−)-elisapterosin B **40**. In each case, the reaction proceeds diastereospecifically to the target as shown in Figure 3.12, box 1. Chida et al.[19] provided synthesis of (−)-agelastatin **45**, which has been shown to inhibit some human cancer cell lines[20] using a domino sigma-tropic rearrangement where the rearranged double bond from the first rearrangement facilitates the second as shown in Figure 3.11, box 1.

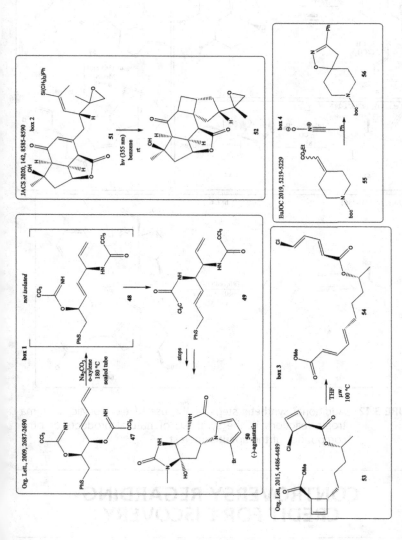

FIGURE 3.11 Sample synthetic steps of the use of electrocyclic or sigmatropic reactions in the synthesis of natural products or other compounds of therapeutic interest.

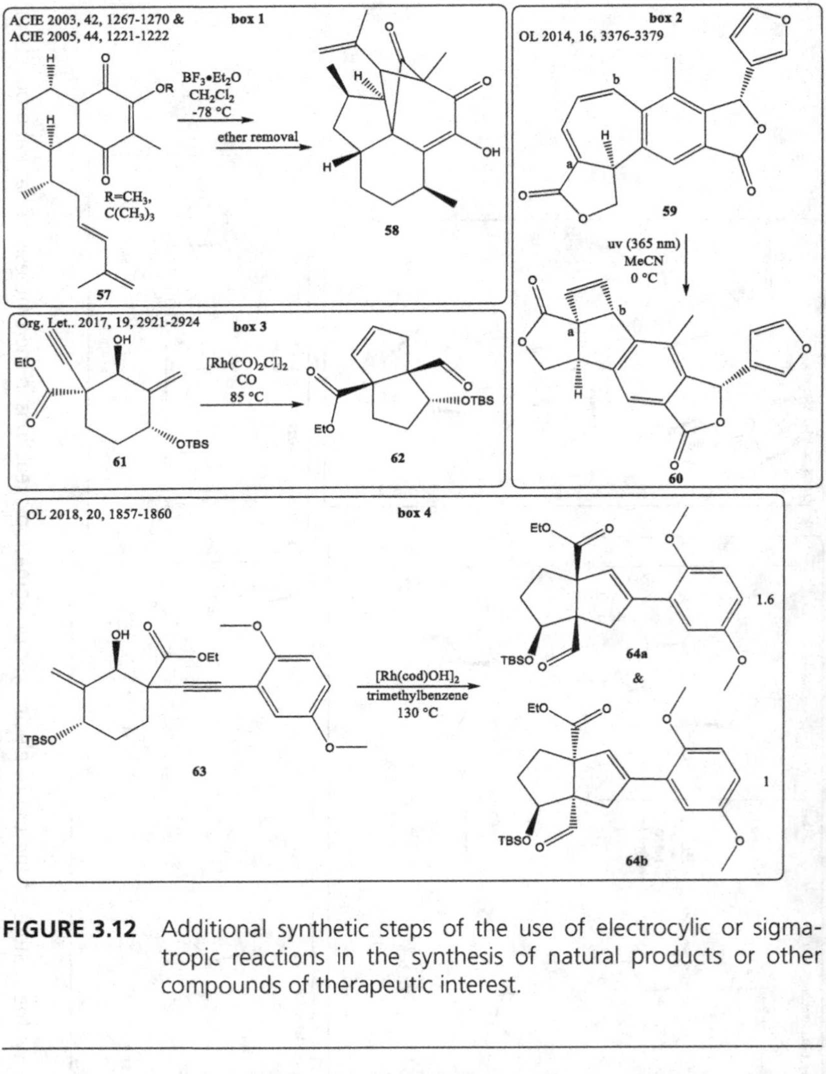

FIGURE 3.12   Additional synthetic steps of the use of electrocylic or sigma-tropic reactions in the synthesis of natural products or other compounds of therapeutic interest.

CONTROVERSY REGARDING CREDIT FOR DISCOVERY

As discussed in Volume 1, a controversy has arisen regarding the credit for Hoffmann's portion of the award. Specifically, E. J. Corey—whose own Nobel Prize is the topic of the next chapter in this volume—has made claims that he should have been given credit for the Woodward–Hoffmann rules as well. This claim is based on a conversation between him and Woodward before the publication of Woodward and Hoffmann's 1965 paper whereby Corey contributed critical insights later used by Woodward as his own. Corey first made these claims publicly after Woodward's death. To date, conclusive evidence to support or refute Corey's claim has not been presented. Jeffrey Seeman is currently authoring a series of articles for *The Chemical Record*[21] that chronicles the history of the development of the now-famous rules.

SUMMARY

The reactions in this chapter were not discovered or even optimized by one person or research team. They are, however, all related to the work done by Fukui and Hoffmann, with (by the time of the award the *late*) Woodward playing a more than slightly helpful role in Hoffmann's contribution; they are after all called the Woodward–Hoffmann rules, not Hoffmann's rules. Had he lived another few years, he probably would have been awarded his second Nobel Prize. What these contributions led to is a deeper understanding of the course of chemical reactions, a better understanding of their mechanisms. With such understanding comes better control over them and the logically corresponding increase in their utility as synthetic processes then followed. As for why it took *decades* for the work to be recognized with a Nobel Prize (remember, the Woodward–Hoffmann paper was published in 1965 and the award was made 16 years later), one can only guess. One logical explanation, however, is that during these intervening years, the utility of these reactions was better harnessed to solve complex and important problems.

GENERAL REFERENCES

https://www.nobelprize.org/uploads/2018/06/fukui-lecture.pdf, last checked 11/9/22.

https://www.nobelprize.org/uploads/2018/06/hoffman-lecture.pdf, last checked 11/9/22.

Smith, Michael B. *Organic Synthesis*, second edition, McGraw Hill, 2002.

Carey, Francis A. & Sundberg, Richard J. *Advanced Organic Chemistry Part A: Structure and Mechanisms*, fourth edition, Kluver Academic/Plenum Publishers, 2000.

Smith, Michael B. *March's Advanced Organic Chemistry*, seventh edition, Wiley, 2013.

NOTES

1 *Oxford Dictionary of Chemistry*, 5th edition, **2004.**
2 https://en.wikipedia.org/wiki/Roald_Hoffmann, last checked 11/11/22.
3 As of this writing, the Ukraine has been defending itself from an invasion by Russia for almost a year. Who knows how and/or if the map will be re-written by the time this war is over.
4 Woodward, R. B.; Hoffmann, R. J. *Amer. Chem. Soc.* **1965**, *87*, 395–397.
5 Nowicki, J. *Molecules*, **2000**, *5*, 1033–1050.
6 Ilardi, E. A.; Stivala, C. E.; Zakarian, A. *Chem. Soc. Rev.* **2009**, *38*, 3133–3148.
7 Wang, Z.; Liu, J. *Beilstein J. Org. Chem.* **2020**, *16*, 3015–3031.
8 Bian, M.; Li, L.; Ding, H. *Synthesis* **2017**, *49*, 4383–4413.
9 Ylijoki, K. E. O.; Stryker, J. M. *Chemical Reviews*, **2013**, *113*, 2244–2266.
10 Hafeman, N. J.; Loskot, S. A.; Reimann, C. E.; Pritchett, B. P.; Virgil, S. C.; Stoltz, B. M. *J. Amer. Chem. Soc.* **2020**, *142*, 8585–8590.
11 Thao, N. P.; Nam, N. H.; Cuong, N. X.; Quang, T. H.; Tung, P. T.; Dat, L. D.; Chae, D.; Kim S.; K, Y.-S.; Kiem, P. V. Minh, C. V.; Kim, Y. H. *Bioorg. Med. Chem. Lett.* **2013**, *23*, 228–231.
12 King, T. A.; Stewart, H. L.; Mortensen, K. T.; North, A. J. P.; Sore, H. F.; Spring, D. R. *Eu. J. Org. Chem.*, **2019**, 5219–5229.
13 Fu, C.; Zhang, Y.; Xuan, J.; Zhu, C.; Wang, B, Ding, H. *Org. Lett.* **2014**, *16*, 3376.
14 Souris, C.; Misale, A.; Chen, Y.; Luparia, M.; Maulide, N. *Org. Lett.* **2015**, *17*, 4486–4489.
15 Zheng, N.; Zhang, L.; Gong, J.; Yang, Z. *Org. Lett.*, **2017**, *19*, 2921–2924.
16 Shao, W.; Huang, J.; Gong, J.; Yang, Z. *Org. Lett,* **2018**, *20*, 1857–1860.
17 Kim, A. I.; Rychnovsky, S. D. *Angew. Chem. Int. Ed.* **2003**, *42*, 1267–1270.
18 Harrowven, D. C.; Pascoe, D. D.; Demurtas, D.; Bourne, H. O. *Angew. Chem. Int. Ed.* **2005**, *44*, 1221–1222.
19 Hama, N.; Matsuda, T.; Sato, T.; Chida, N. *Org. Lett.* **2009**, *11*, 2687–2690.
20 a. D'Ambrosio, M.; Guerriero, A. Debitus, C.; Ribes, O.; Pusset, J.; Leroy, S.; Pietra, F. *J. Chem. Soc. Chem Commun.* **1993**, 1305–1306. b. D'Ambrosio, M.; Guerriero, A.; Pietra, F.; Ripamonti, M.; Debitus, C.; Walkedre, J. *Helv. Chim. Acta.* **1996**, *79*, 727–735, c. Mason, C. K.; McFarlane, S.; Johnston, P. G.; Crowe, P.; Erwin, P. J.; Domostoj, M. M.; Campbell, F. C.; Manaviazar, S.; Hale, K. J.; El-Tanani, M. *Mol. Canc. Ther.* **2008**, *7*, 548–558.
21 https://onlinelibrary.wiley.com/doi/toc/10.1002/(ISSN)1528–0691.woodward-hoffmann-rules, last checked 11/11/22.

1990—Corey

4

Elias James (E.J.) Corey won the 1990 Nobel Prize in chemistry "for his development of the theory and methodology of organic synthesis." While to the uninitiated reader, this may seem overly simplistic or basic; it is anything but. For sure, Corey's work simplifies the task of the multistep synthesis of a complex chemical target. This systematic way of planning out the elaboration of simple starting materials to evermore complex building blocks to an ultimate target has made the entire field more accessible and furthered not just chemical synthesis but also (even if only by extension) the pharmaceutical industry. It does so by easing the chemical preparation of compounds of human interest or importance. Although Corey was recognized for this work vis-à-vis a Nobel Prize in 1990, his development of the method, even computer-assisted retrosynthetic analyses, dates back further.[1]

In his Nobel lecture, Corey tells a humorous story about his days as a student while taking a course on advanced synthetic organic chemistry in 1947. During this course, the professor commented that since so few important reactions (five of them) had been discovered in the last 50 years, very few synthetic methods remained to be found. Today, this is laughable though, perhaps, reactions **can** be broken down into six broad categories: 1) addition reactions, 2) substitution reactions, 3) elimination reactions, 4) rearrangements, 5) acid–base reactions (i.e., proton exchanges), and 6) metal exchanges, the last of which can justifiably be considered a specific type of substitution reaction. It should also be noted that only 1–4 are typically used to move a synthetic sequence forward. The volume of synthetic methods is almost immeasurable and still growing. At this point, it is only with the systematic approach developed by Corey that complex molecules can be rationally synthesized.

According to Corey, even into the 1970s, a widespread systematic problem-solving approach had not yet been applied to the field of synthesis. Even as early as the mid-1960s, the approach that is now referred to as retrosynthetic (or antithetic) analysis was picking up steam. Prior to this time, syntheses were undertaken by choosing some sort of starting material deemed appropriate, whether for structural similarities, established stereochemistry, or a myriad of

DOI: 10.1201/9781003006848-4

other reasons. A series of chemical reactions then followed that step-by-step enabled chemical transformations that furnished the final product.

Retrosynthetic analysis—or simply, retrosynthesis—is a thought exercise that starts at the product and works backward to break down—or disconnect—the target into increasingly simpler targets. It should be obvious that the transformation of these simpler targets to the more complex ones could be brought about using known chemistry; if there are no known chemical reactions capable of joining the two fragments or performing the chemical change, such would not be part of any logical plan. The plan must rely on known chemistry or else the synthesis will dead-end. Eventually, after enough disconnections, commercially available or otherwise easily obtained starting materials are arrived at. The benefits of this sort of approach were recognized very quickly. For example, it usually makes incorporating minor changes into the target structure trivial to execute. This may not seem important, but since many products and/or pharmaceutical compounds have such small changes between them, an entire library of compounds can often be made using one general plan. Also, having a set of followable and more importantly *programmable* guidelines opened the doors to computer-assisted synthetic planning. This has only increased in recent years with the rapidly increasing computer power and storage and the power of artificial intelligence. These days, online databases such as CAS SciFinder have access to millions of individual reports of chemical reactions, all of which are probed in every search to the user's

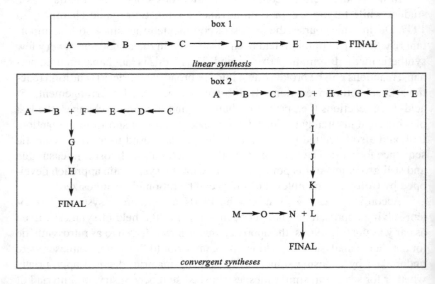

FIGURE 4.1   Examples of linear and convergent syntheses.

query. This is only going to increase in the coming years; we are unlikely to be near the ceiling of this power. The sophistication in sorting and analyzing the results—the hits—returned by such searches is also increasing, making it even easier for the user to find data applicable to their specific problem. Finally (and most importantly to me), such a system is *teachable*. This last point means that anyone with sufficient motivation can learn how to do this.

Finally, before moving on to discussing synthetic planning, one last introductory comment must be made about multistep syntheses. Generally, a synthetic plan will either be what is called a linear (Figure 4.1, box 1) or convergent synthesis (Figure 4.1, box 2), either of which can be the product of the kind of retrosynthetic analysis described. A linear sequence proceeds along a series of targets in a straight "road" of sorts, going from starting material to the final product. A convergent synthesis, on the other hand, has more than one linear sequence meet to combine sophisticated intermediates along the way.

RETROSYNTHETIC ANALYSIS

At their most simplistic, the guidelines (rules is probably too strong a word) direct the chemist to focus on certain bonds within a molecule as ideal to make during the synthesis. Often, these are bonds that either directly involve a functional group or are within a certain proximity of a functional group. Some functional groups allow chemistry to occur at a greater number of atoms away than others with combinations/sequential functional groups doing this even more so. Other times, stereocenters, rings, or other structural features figure more prominently. Since Corey's development of this approach, it is all but impossible to read an article in which a multistep synthesis of a complex target is reported and not see the *retrosynthesis*. This retrosynthesis or retrosynthetic (or antithetic) analysis is essentially the application of Corey's guidelines. If nothing else does so, that *everyone* uses it should indicate to anyone the importance of the approach. The process takes the molecule apart bit-by-bit, breaking it into smaller and simpler pieces; pieces you (the synthetic chemist) know you can combine to arrive at as the target or one of the intermediates previously shown in the literature to lead to the target after some number of chemical reactions. The result of this analysis is nothing less than a roadmap to use in the synthesis of the target; the result is a synthetic plan. In truth, like any other ***plan*** in life, adjustments are made when the plan is executed, and unexpected problems pop up. This, of course, makes the plan no less important. It is also worth noting that this approach should furnish multiple plans for the same target. The chemist then must choose the most appropriate one keeping in mind many factors including safety and both the cost and availability of starting materials/other reagents. It is also possible that more than one plan may share some number of intermediates. In many—though not all—instances, the novelty of the plan should also be considered. The weight attributed to novelty will vary, though, depending on the goals of the project. In cases where the novelty may mean less, research teams are keen to showcase methodology developed in their lab and will plan a synthesis around including it.

Around the middle of the 20th century, the complexity and sophistication of chemical synthesis dramatically increased. Corey attributes this to five stimuli.

1. The formulation and understanding of detailed electronic mechanisms for the fundamental organic reactions
2. The introductions of conformational analysis of organic structures and transition states based on stereochemical principles
3. The development of spectroscopic and other physical methods of structural analysis
4. The use of chromatographic methods of analysis and separation
5. The discovery of new selective chemical reagents

How an understanding of mechanisms helps synthesis

It is fair to wonder how these stimuli impacted a field as wide and complex as organic synthesis. Of the options, the determination of what is called the reaction mechanism is one of the most important. To understand why, it must first be understood what is meant by a reaction mechanisms. Although a discussion of how we depict them is inappropriate here, they can most briefly be described as a molecular, even atomic, level accounting for the formation and rupture of every bond as an individual starting material or intermediate is transformed into the next intermediate or product. Understanding how a chemical reaction proceeds on this level permits the tweaking of chemical reagents and/or reaction conditions that optimize the formation of a desired product and often the concurrent suppression of undesired byproducts.

How spectroscopy helps synthesis

Spectroscopic analysis and other physical analysis methods allow for the rapid confirmation of a molecule's identity. Prior to the expansion of both—especially since the development of nuclear magnetic resonance (NMR) into a virtually ubiquitous analysis method—chemical analysis, whereby compounds were derivatized to other chemical compounds was commonly used to confirm chemical structures and/or transformations. In cases where the compound being made is a known compound, its melting point (if it is a solid) would be compared to the known value to aid in identification. Spectroscopy of all kinds takes far less time and is almost always more accurate. Furthermore, some spectroscopic analysis methods such as NMR do not destroy the material; you can recover the sample you analyzed unchanged. In short, these methods of analysis allowed for the more rapid and more reliable identification of the products of chemical reactions. Knowing structures sooner and with better confidence made synthesis more efficient and no doubt helped ascertain the mechanisms of many reactions since the techniques could just as easily be applied to solving the structure of byproducts and even intermediates.

How chromatography helps synthesis

Chromatography is one of many methods of purification. Although—since it is very resource-intensive—it is not commercially used at the multi-kilogram scale needed in the synthesis of the enormous quantities of materials produced by the pharmaceutical industry, amongst others, is perfectly appropriate on the smaller scale. The smaller scale is precisely the scale ordinarily encountered in synthetic labs; placing chromatography in position to dramatically impact the field of synthesis. Modern instrumentation has automated chromatographic purifications, making any individual researcher more productive when using it since it enables them to do other work in the meantime. It is also much more broadly applicable to the purification of chemical compounds compared with recrystallization and distillation. Compounds that are not solids cannot be recrystallized, for example. On the other hand, compounds whose boiling point cannot be easily accessed, even with vacuum distillation, cannot be purified by distillation. Generally speaking, only very volatile compounds cannot be purified by chromatographic techniques, and even this limitation is more closely due to the fact that such compounds are difficult or impossible to separate from the solvents used in chromatography due to the compound's volatility, not a failure of the method itself. There are also chromatographic methods that even permit the separation of enantiomers, a separation difficult, at best, via recrystallization and impossible with distillation. All of this can be simplified to say that chromatography allows for the purification of a wider range of substances than other methods. If your substances are more pure, subsequent chemical reactions often—and not surprisingly—work better. Although there are certainly exceptions, using incompletely purified compounds may cause side reactions, destroy reagents, or both in any subsequent reactions.

How conformational analysis helps synthesis

The incorporation of conformational analysis has its biggest impact on the selective formation of stereoisomers in the product mixture. This point serves as somewhat of a bridge between the next point and the point about mechanisms. This is because the incorporation of conformational analysis is critically dependent on the mechanisms of the reaction. Combining the two leads to the selective generation of stereoisomers, the larger percentage of a mixture a desired product is, the easier both subsequent synthetic steps and purification will be. Conformational analysis can also be invoked during intramolecular reactions from the point of view of a reaction happening and being prevented.

How selective chemical reagents help synthesis

This is often referred to as chemoselectivity (Figure 4.2, box 1), selectivity regarding the chemical reaction that occurs. Regiochemistry (Figure 4.2, box 2) and selectivity is similar and entails which product is favored when multiple products of the same general reaction type can be formed.

Even a novice should be able to immediately appreciate that chemical conditions that will react with only one functional group in the presence of others or give only one product when multiple options are possible are highly desirable. Such allows for much more efficient access to the final target. Even in the years since Corey's 1990 Nobel award, never mind since his professor's now comical comments in 1947, much progress has been made in this arena. Nevertheless, sometimes, known (or maybe…yet undiscovered selective) reactions coincide with unfavorable combinations of functional groups in a target, conspiring against a synthetic chemist.

Fortunately, even in these situations, not all is lost, nowhere close in fact and it is arguably in these cases that the retrosynthetic plan is most important. To overcome such competing reactivity problems, what are called

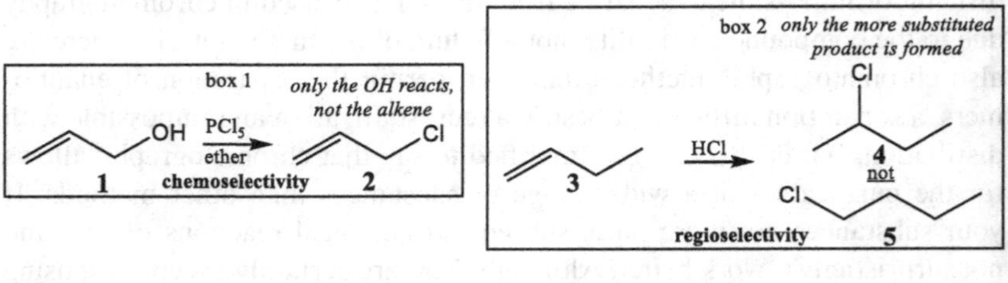

FIGURE 4.2   Comparing chemoselectivity and regioselectivity.

protecting groups are typically employed; these can just as logically be thought of as masking groups. In short, such groups convert the offending part of the molecule—the part of the molecule causing the *undesired* reaction—into a different functional group until any and all reactivity this functional group would engage with is done. At this point, the previously mentioned conversion is undone, revealing the original group again. Naturally, the protecting group **must not** react with any of the reaction conditions employed during the intervening steps. Furthermore, incorporating these protecting groups into the molecule **must not** also react with other functional groups present. Finally, removing the protecting group **must not** react with other functional groups. Fortunately, so common is this problem in multistep syntheses that an enormous number of options for protecting groups exist for almost every functional group imaginable. Another aspect of these important synthetic tools, which must be mentioned, is that both the reaction that puts the protecting group on and the one that takes it off must be very high yielding to avoid waste of material. Although this is always the case for chemical reactions, it is particularly important with protecting group reactions since these steps do not move forward the building of the target. This makes loss of material more wasteful in these steps since the synthesis is not moved forward during these steps. As a result, avoiding the use of protecting group is very desirable, so desirable in fact that when a synthesis is reported in the peer-reviewed literature that avoids protecting groups, the authors often brag about this in the title of the manuscript.

Recognize that every protecting group adds two steps to the synthetic sequence: 1) putting the group on and 2) taking it off. Efficiencies, however, are possible with a good retrosynthetic plan. With appropriate planning, the deliberate removal of multiple protecting groups at the same time can be incorporated into the plan. Sometimes, functionalities in the molecule even permit the use of a functional group in one part of the molecule to react with another functional group in a different part of the molecule to essentially use each to protect the other. A solid and well-thought-out plan also unlocks yet more complex possibilities. For example, sometimes, use of a starting material that contains what is literally or at least effectively amounts to a protecting group is possible. This saves the user at least one step in their chemical sequence. Alternatively, the plan may make clear an alternative order in which the reactions can be executed, obviating the need for one or more protecting groups. The best of the best are even capable of greater insights. For example, protecting groups such as an allyl group can be used to protect an alcohol with the full intention of revealing the alcohol (i.e., removal of the protecting group) using some manner of sigmatropic rearrangement that at once serves as this deprotection and executes an altogether different and necessary chemical transformation. In instances where the deprotected group is meant to react with other functional groups in the molecule, the timing of deprotection can be planned to be subsequent to the incorporation of this other group to enable this transformation to occur concurrently with the deprotection step, if the chemical conditions for the transformations are compatible or better yet, identical to the conditions used for the deprotection.

EXECUTING RETROSYNTHETIC ANALYSIS

How, though, is retrosynthesis done? At its most basic and simplistic level, one can look at a molecule, identify one single transformation they know to execute to synthesize the target, and make the starting material for that reaction be their new target to analyze. In the same way, this is arguably a basic explanation of the topological strategy described later. Although this approach can work, especially for simpler molecules, a more logical approach exists. For highly complex molecules, Corey describes several different strategies:

1. Transform-based strategies
2. Structure-goal strategies
3. Stereochemical strategies
4. Functional group strategies
5. Topological strategies

Each approach has its merits and shortcomings, and a one-size-fits-all approach should not be employed. In fact, an approach that incorporates all the relevant[2] strategies should be considered. Here, barely the basics of each approach will be discussed. The space available here is woefully insufficient to provide anything more than what is currently involved in this chapter. For a more extensive discussion and a plethora of examples, it is likely best to go to the source: E.J. Corey himself and his epic missive "The Logic of Chemical Syntheses."[3]

TRANSFORM-GUIDED RETROSYNTHETIC STRATEGIES

The key tenet of this approach can be described as simplifying the molecule to a target that has certain keying features, for example, some sort of ring or polycyclic structure or a complex array of stereocenters. In these sorts of situations, disconnections that build the molecule off such rings or employ reactions that are stereoselective or better yet specific are logical. Starting with materials that have in their structures the established stereochemistry is also strategic. It is common for a starting material with elaborate stereochemistry to be a natural product or at least a minimally modified natural product.

STRUCTURE-GOAL STRATEGIES

In its most general sense, this approach focuses on identifying a specific starting structure that contains some sort of structural aspect—be it connective or stereochemical. This sort of approach comes with a trade-off. While on the one hand it serves to address a specific challenge that has been identified, it also has the effect of narrowing the options within the synthetic pathway. This is because now, certain functional groups may be already present or necessary later for the synthesis. Arguably, a similar effect is observed when particular research groups have themes or certain chemical reactions they have invented and want to demonstrate the utility and scope of. In this sort of case, the synthetic pathway would be deliberately designed to showcase such a particular chemical reaction. Although undoubtedly important to invent new reactions and show their scope, this sort of deliberate incorporation of such a reaction does not always produce the most efficient pathway to the final complex target though it does expand the synthetic toolbox.

STEREOCHEMICAL STRATEGIES

The goal of stereochemical strategies is to reduce the stereochemical complexity during the retrosynthetic analysis. This is done by the elimination of stereocenters in a target molecule. Naturally, only stereocenters that can be synthetically set later (or provided by commercially available materials) should be removed during the retrosynthesis. Removing stereocenters that cannot be so set later will dramatically reduce the yield of later steps since a more complicated mixture of products will result. Alternatively, if a stereocenter in an intermediate is destined to be destroyed in later steps, oftentimes, the synthetic plan would not concern itself with this stereocenter unless it has a deleterious effect on the stereochemical outcome of other reactions due to conformational effects.

FUNCTIONAL GROUP-BASED STRATEGIES AND OTHER STRATEGIES

Certain functional groups, because of the breadth and value of chemical transformations they permit, hold a position of higher priority. These include the olefin, carbonyl, acetylenic, hydroxyl, carboxyl, amino, nitro, and cyano groups. Meanwhile, the azo, disulfide, and phosphine groups are less versatile and only crucial when present in the target. Finally, the halides, selenoxide, phosphorus, sulfone, trimethyl silyl, and borane are considered peripheral. These are shown in Figure 4.3.

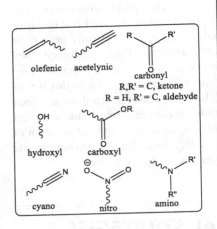

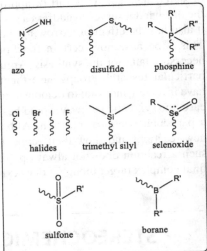

FIGURE 4.3 Functional groups important in synthetic planning.

TOPOLOGICAL STRATEGIES

While a strong grasp and vast knowledge of synthetic reactions are without question essential to both synthesis and retrosynthesis, there is much more reason behind choosing disconnections over others. For example, in the simple structure in Figure 4.4, only disconnections a-e make sense while f does not.

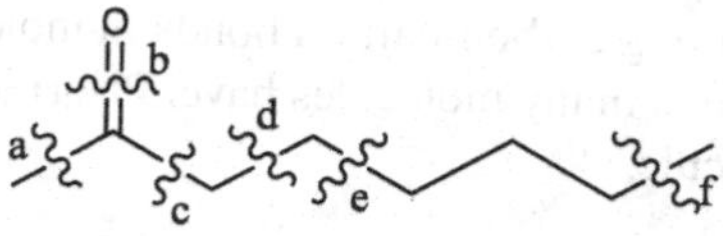

FIGURE 4.4 Examples of logical and illogical disconnections.

What is important to notice is the relative proximity of the disconnection to the functional group. Specifically, the disconnection should not be more than three bonds away from a functional group. As new chemical reactions are developed, this general guideline may change, but it serves as a good general guideline. Moreover, this "rule of three" does not universally apply to every functional group; at least not directly and efficiently. In the example here, disconnection b involves the functional group and can be considered zero bonds away. Meanwhile, a and c disconnect an atom from the functional group bearing carbon atom; we will call this one bond away. Disconnection d, meanwhile, would be called two bonds away, and e, three bonds away.

DISCONNECTION BY THE NUMBERS

A sophisticated approach such as Corey's is likely too complicated for a novice just learning organic chemistry. When introducing the topic of synthesis to such learners, a more simplistic approach is necessary. As an analog to Corey's approach, Smith devised a retrosynthetic analysis approach called *Disconnect by the Numbers*[4] that can be applied to relatively simple structures encountered in a typical two-semester undergraduate organic chemistry course sequence or early graduate school. Although less attractive for the application to the synthesis of very complex products, as a tool for teaching the problem-solving approach used in a multistep synthesis, the approach has enormous value. The process focuses on a single criterion: the relative ability to make the disconnected bond via a chemical reaction. Smith's protocol assigns priorities to each bond based on carbon-carbon bond forming reactions that are usually covered in a two-semester course sequence on organic chemistry. So, by its nature, it omits a wide range of sophisticated reactions, and with good reason, this is meant to be a basic introduction to complex structures for novices making the transition to advanced. Smith's approach includes tables of data and reactions to guide the user in making decisions based on the scores assigned. Tables also help assign these numbers based on the importance of forming carbon-carbon bonds in molecules with polarized functional groups, which many molecules have. Priorities are based on a few key regions in a molecule:

1. Bonds connected to the functional group
2. Bonds second from the functional group
3. Bonds third from the functional group
4. Stereocenters

All bonds get assigned a priority and the bonds with the highest priority are the ones disconnected. Smith also provides a table of synthetic equivalents that assist in converting the disconnected molecular pieces into "real starting materials." Critical to this is deciding which piece will be the donor and which will be the acceptor. The donor is the piece that will function as the Lewis base (electron donor), and the accepter would be a Lewis acid (electron acceptor).

MANAGING A LARGE VOLUME OF CHEMICAL INFORMATION

One of the keys to successfully implementing **any** retrosynthetic strategy is a command or easy access to a list of chemical transformations. In my opinion, one of the best ways to acquire this command is to think of every chemical reaction in two ways: 1) reactions that make a product and 2) reactions of the class of compounds but specifically the functional group undergoing the transformation. Control of the former should be obvious; you are trying to prepare something, having a broad knowledge of the ways to do a wide array of chemical transformations is essential to doing this. For any standalone chemical transformation, this is perfectly sufficient. However, for a multistep sequence of reactions being used to synthesize a complex target, far more is necessary; just a knowledge of how to make a product is insufficient. This is because many, most actually, compounds have more than one functional group. For the synthesis of these compounds, one must also have a command of the reactions from the point of view of the functional groups undergoing a chemical reaction. This finer detail is a major driving force for the order of chemical reactions during a multistep synthesis since it enables vision into how an entire molecule may react. When avoiding competing/conflicting reactivity by changing the order of chemical reactions is not possible or practical, the researcher is left with two options: 1) invent the chemistry they need or 2) use protecting groups. Whatever the strategy used to overcome the challenge, only a grasp of the reactivity at the level described here will get the job done.

Such an organization is relatively easy to incorporate into even a two-semester sophomore-level course sequence. If reactions are deliberately presented to students from both points of view, there would be a natural buildup of all reactivity at every step. Such an approach ought to better prepare students for the complexity that awaits them should they choose to continue pursuing organic chemistry.

SELECT EXAMPLES OF RETROSYNTHESIS IN THE PEER-REVIEWED LITERATURE

The balance of this chapter presents a variety of retrosynthesis analyses as they are found in a peer-reviewed publication. Even a brief glance at papers discussing total synthesis reveals a wide range of retrosynthetic analysis depictions. I suspect, more than anything scientific, this is due to the constraints of print space in publications. In reality, the synthetic plan is far more elaborate and specific than the space in most publications allows for. Authors of papers usually choose to focus their published retrosynthesis on key reactions or reactions their lab has developed and are consequently found desirable to be so showcased. As a result of these limitations, looks can be somewhat deceiving. An inexperienced reader may take from looking at such retrosynthetic analyses that the corresponding syntheses are very short. The actual number of steps used to execute the synthesis is almost always far greater. Remember, there are space limitations that demand at least some manner of abbreviated presentation. Also, there are syntheses of natural products in excess of 50 individual steps. Listing out the plan for every step is excessive and largely unnecessary. Even the actual, that is, forward, synthesis is typically shown in abbreviated form by detailing the reagents for multiple steps above each arrow. It is easy to make the argument that the actual synthesis is "more important" than the plan; if even the actual synthesis is abbreviated, ought not the plan be abbreviated too?

Furthermore, the retrosynthesis often omits the preparation of some of the starting materials, even when they are not commercially available. Sometimes this is because they are commercially available, meaning that the chemists reporting this work do not have to make it. Other times, it is because the synthetic protocols for making the compounds have already been reported and are not the subject of the current work. In essence, the retrosynthesis often focuses on the stage or steps in a synthesis that is central to or novel within the current work.

Also, more often than many probably want to admit, the elegantly presented plan in a publication or oral presentation is not always the original plan. As there is currently no *Journal of Failed Research*,[5] there is often only enough space for the final successful plan. It can certainly be argued that such aborted plans may be informative and, therefore, could justifiably be published in the supplemental information (which is not printed and, thus, does not "take up precious space") or even simply stored online as a webpage. Unfortunately, there does not appear to me to be any serious call for this sort of disclosure despite the reality that someone else may pursue a path found by another to fail as a result.

In Coster and Magolan's[6] synthesis of (+)-angelmarin (Figure 4.5), very few meaningful details are provided, particularly to a novice reader. For example, the first transform/disconnection from **7** to **8** is accomplished through some manner of esterification reaction. There is a variety of ways to accomplish this transformation and the retrosynthetic analysis gives no hint at all of which is used. In fairness, such specific detail would *never* be part of a retrosynthetic analysis. The next move is described as simply 5-exo-tet. This likely only makes sense to a more experienced/learned organic chemist who no doubt will recognize this as a type of ring closure. The Shi epoxidation likewise is sure to only make sense if you have learned that such a reaction is the asymmetric (stereoselective or stereospecific) epoxidation of an alkene. Finally, an alternative format altogether is used to show that the alkene will be generated using a cross-metathesis reaction which itself requires a different alkene or an alkyne.

Moslin and Jamison[7] provide a little more detail in their retrosynthetic approach to the synthesis of (+)-acutiphycin (Figure 4.6). This level of detail

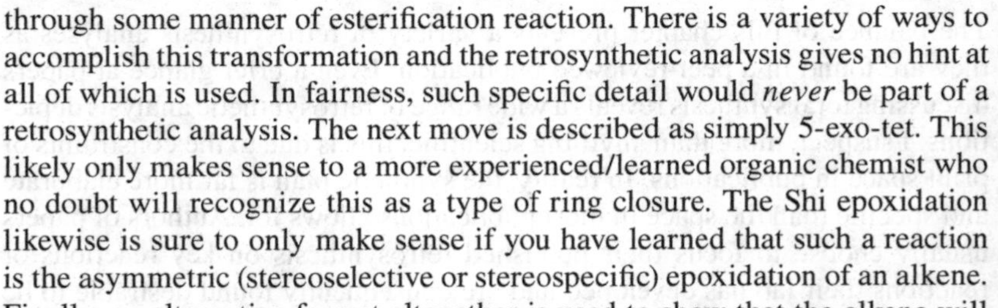

J. Org. Chem. **2009**, 5083

FIGURE 4.5   Coster and Magolan's retrosynthesis of (+)-angelmarin.

is far more informative though simultaneously fails to directly say something about the *order* of the reactions. The plan even goes on to specify the identity of protecting groups that will be used. It also shows multiple transforms/ disconnections at the same time to conserve space, even though these reactions are not actually done simultaneously.

Dyker and Hildebrandt[8] used a transition metal-catalyzed domino process that includes an intramolecular Diels-Alder reaction in their synthesis of heliophenanthrone. For a reader hungry for details, the reported retrosynthesis (Figure 4.7) is maddeningly bereft of meaningful details. Their model system (Figure 4.8) and recognizing that such a domino process would furnish heliophenanthrone with the appropriate substitutions on the starting material enabled them to identify appropriate starting materials using retrosynthesis.

Todano et al.[9] reported their synthesis of clavilactones in 2018. At the conclusion of their synthesis, they discovered that the original structural assignment of one, clavilactone D[10] was incorrect. Fortunately (for them), the synthetic plan (Figure 4.9) they had was flexible enough to be followed with a variety of analogs to permit the synthesis of a range of derivatives. This was particularly possible because of their choice of metathesis reactions, which are often functional group tolerant for key steps. This is not simply good fortune but also good planning.

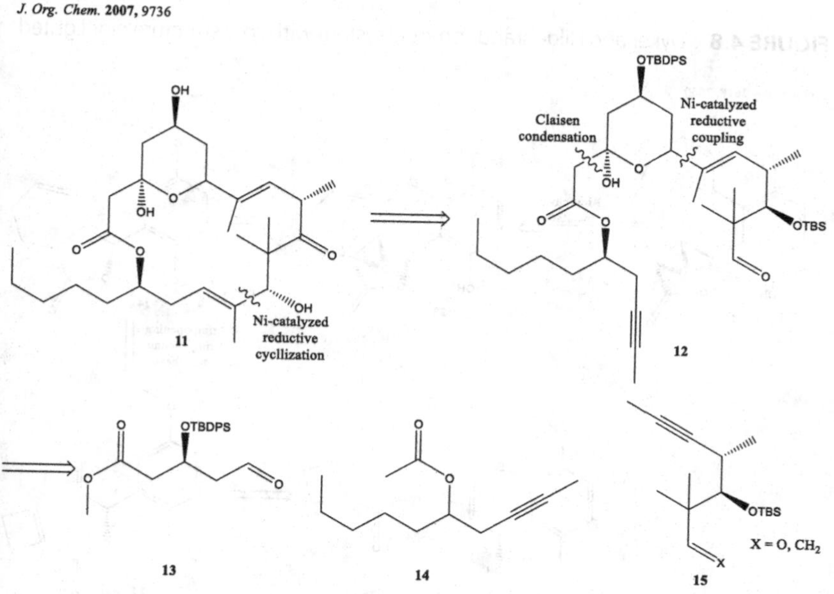

J. Org. Chem. **2007**, 9736

FIGURE 4.6 Moslin and Jamison's retrosynthetic approach to the synthesis of (+)-acutiphycin.

J. Org. Chem **2005**, 6093

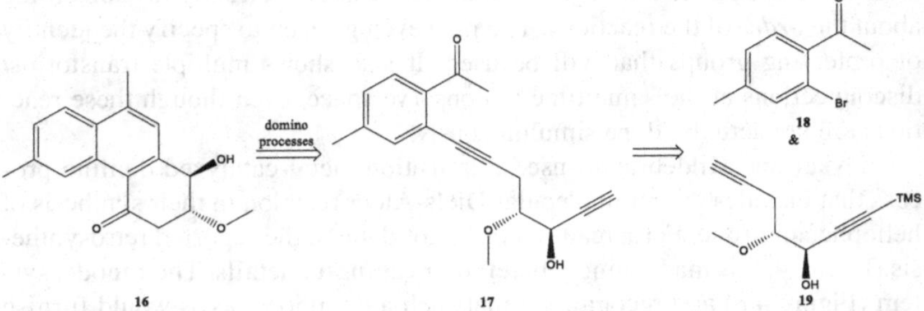

FIGURE 4.7 Dyker and Hildebrandt's retrosynthesis of heliophenanthrane.

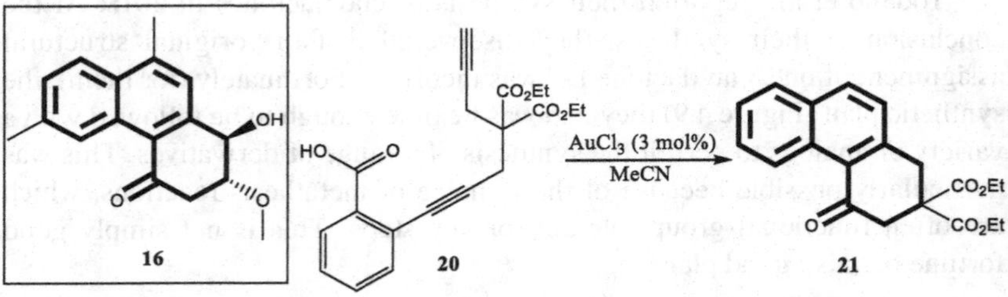

FIGURE 4.8 Dyker and Hildebrandt's model system with core structure highlighted.

J. Org. Chem **2018**, 7060

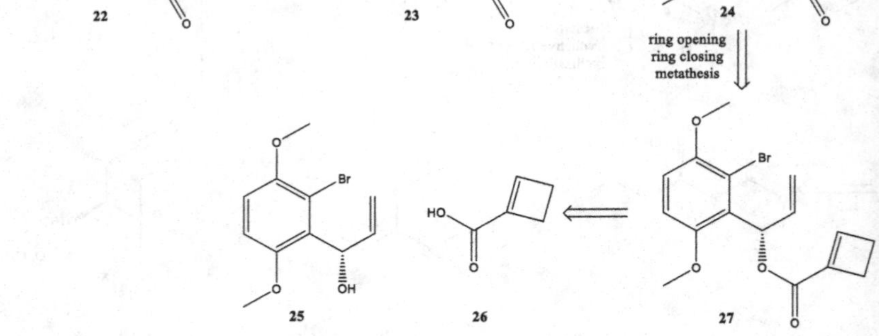

FIGURE 4.9 Todano et al.'s retrosynthesis of clavilactones.

SUMMARY

All these sorts of sophisticated synthetic approaches are possible at least in part due to the synthetic chemist's creative and reaction genius. However, it is hard to envision a way that such genius could be used to the sophisticated ends of complex chemical synthesis without Corey's work. In my opinion, Corey's work is different from the other Chemistry Nobels that relate to organic synthesis. It can easily be considered the pinnacle of an intellectual crescendo that started with Baeyer and Wallach—who expanded the utility of organic chemistry—to Robinson and especially Woodward, the latter's brilliant syntheses were nothing less than a coming of age of organic synthesis as a tool to prepare compounds of enormous complexity. Corey's work can be considered to be one of *teaching* every subsequent generation of organic chemists how to devise a synthetic plan for preparing such complicated targets. This by no means is to minimize what Corey did. Quite the opposite is my intention. Corey articulated guidelines that allow any learned organic chemist to design a synthetic plan for a complex target. The guidelines quite literally enable complex syntheses to be designed by all sufficiently learned organic chemists; it dramatically increases the number of chemists able to perform this sort of research. It is possible that no single contribution to the field has ever done more to further it.

GENERAL REFERENCES

Corey, E. J. & Cheng, X.-M. *The Logic of Chemical Synthesis*, John Wiley, and Sons, 1995.

Smith, Michael B. *Organic Synthesis*, second edition, McGaw Hill, 2002.

https://www.nobelprize.org/uploads/2018/06/corey-lecture.pdf, last checked 11/21/22.

NOTES

1 Corey, E. J.; Wipke, W. T. *Science* **1969**, *166*, 178–192. Corey, E. J.; Long, A. K.; Rubenstein, S. D. *Science* **1985**, *228*, 408–418. Corey, E. J. *Quart. Rev. Chem. Soc.* **1971**, *25*, 455–482.

2 For example, a target lacking stereocenters or that is achiral would not consider stereochemical strategies.

3 Corey, E. J.; Cheng, X.-M. *The Logic of Chemical Synthesis*, **1995**, John Wiley, and Sons.

4 Smith, M. B. *J. Chem. Ed.* **1990**, *67*, 848–856.

5 Much to the chagrin of many.

6 Magolan, J.; Coster, M. J. *J. Org. Chem.* **2009**, *74*, 5083–5086.

7 Moslin, R. M.; Jamison, T. F. *J. Org. Chem.* **2007**, *72*, 9736–9745.

8 Dyker, G.; Hildebrandt, D. *J. Org. Chem* **2005**, *70*, 6093–6096.

9 Takao, K.; Mori, K.; Kasuga, K.; Nanamiya, R.; Namba, A.; Fukushima, Y.; Nemoto, R.; Mogi, T.; Yasui, H.; Ogura, A.; Yoshida, K.; Tadano, K. *J. Org. Chem* **2018**, *83*, 7060–7075.

10 Arnone, A.; Cardillo, R.; Meille, S. V.; Nasini, G.; Tolazzi, M. J. *J. Chem. Soc. Perkin Trans I* **1994**, 2165–2168.

Index

Note: **Bold** page numbers refer to tables; *italic* page numbers refer to figures.